D1068087

Gerhard G. Habermehl

Venomous Animals and Their Toxins

With 44 Figures and 44 Tables

Springer-Verlag
Berlin Heidelberg New York 1981

Prof. Dr. rer. nat. Gerhard G. Habermehl

Direktor des Chemischen Institutes
Tierärztliche Hochschule Hannover
Bischofsholer Damm 15
D-3000 Hannover 1

Based on the translation of the 2nd German edition:
"G. Habermehl, Gift-Tiere und ihre Waffen", Springer-Verlag
1977, ISBN 3-540-08461-4

ISBN 3-540-10780-0 Springer-Verlag Berlin Heidelberg New York
ISBN 0-387-10780-0 Springer-Verlag New York Heidelberg Berlin

Library of Congress Cataloging in Publication Data. Habermehl, Gerhard. Venomous animals and their toxins. Enl. translation of Gift-Tiere und ihre Waffen, 2nd ed., 1977. Includes bibliographical references and index. 1. Poisonous animals. 2. Venom. I. Title. QL100.H2713 1981. 615.9'42. 81-8866. ISBN 0-387-10780-0 (U. S.). AACR2

Typesetting and binding: G. Appl, Wemding. Printing: aprinta, Wemding.
2152/3140-543210

Preface

Venomous Animals have been a threat to man at all times, in the warm and wilder regions more than in the temperate areas. People in especially dangerous regions know about these risks and live accordingly. However, with modern tourism and nearly unlimited travel opportunities more and more people without experience and knowledge about venomous animals come into contact with them; this book is intended to provide these people with an introduction to the subject.

Venomous animals, their habits, their whole ecology and their venoms have been the object or research since the beginning of this century; truly intensive work, however, first started about thirty years ago. Medical treatment therefore has been changed by new insights in the mechanism of action and the constituents of the various venoms. In this regard this book is also directed to physicians, biologists and chemists to give them an introduction in this important and interesting field. New aspects of treatment of envenomations are reported. This book cannot replace bigger textbooks and monographs – they are cited in the references – but it gives an overview and an entry into this field. The original German edition was written at the request of colleagues and students of medicine, biology and chemistry as well as frequent travellers in tropical countries. It was sold out within eight months, thus showing that it really filled a gap. Many colleagues and friends outside Europe asked for an English version, thus encouraging Springer-Verlag to publish it.

I have to thank quite a number of colleagues for discussions, suggestions and reading the manuscript. They are Professor Dr. D. Magnus, Darmstadt and Professor F. Kornalik M. D. Prague, Professor Dr. T. Freyvogel, Basle, Professor Dr. H. Alistair Reid OBE, Liverpool, Dr. J. W. Daly, N. I. H., Bethesda, Md., USA, Dr. Margaret Weiss, Bethesda and Dr. D. Mebs, Frankfurt University Hospital.

I owe special thanks to my good friend Professor Findlay E. Russel, M. D., Ph. D., University of Arizona, Tucson, who was helpful in many, many respects; without him this book would not have appeared. Thanks are also due to the staff of Springer-Verlag, especially Dr. F. L. Boschke, Heidelberg and Dr. Licker, New York, who contributed much to the style and make-up of this book; the cooperation with them has been a pleasure.

Hannover, May 1981 G. G. Habermehl

Table of Contents

Introduction

Since antiquity, venoms from plants and animals have not only posed a danger to humans but also a challenge to our ingenuity. There are numerous, striking, often bizarre examples of this from diverse eras and geographic areas. Indians in Central America and in the Northwestern parts of South America today still use the skin secretions of dendrobatid frogs for hunting as arrow poison. In medieval times dried and powdered Spanish flies were (mistakenly) used as a sexual stimulant; and in China and Japan for 4000 years dried and powdered skins of toads have been used as heart medicines (Ch' an-su and Sen-so). And in the Old World, parallel to that, extracts of *Scilla maritima* (sea-onion) or *Digitalis* (foxglove) were used for the same purpose. Modern science was able to show the chemical similarity of both substances.

Although humans certainly have been aware of the poisonous nature of some snakes from the beginnings of our history, it took a long time until ideas about their venoms took form. Mithridates, king of Pontus (123–63 B. C.) drank the blood of snakes to become immunized against snake bite. Maimonides in 1198 wrote a treatise on "Poisons and Their Antidotes" in which snake bite envenomation is mentioned, but there is nothing told about the origin of snake venom. The Italian physician Francesco Redi (1626–1697) in 1664 wrote a book "De Venenis Animalibus" and was the first to describe the venom apparatus of snakes. He showed that venom could be obtained only from the venom glands and teeth and not the whole animal which until then was supposed to be toxic. A similar story may be told about the salamander. In ancient times the most unusual ideas were reported. Plinius Secundus, the famous Roman author of a multivolume book "Historia Naturalis", Natural History, wrote among others: "Inter omnis venenata, salamandra scelus maximum est" – "Among all venomous animals, the salamander is the most wicked one! Other animals harm individuals only and do not kill several at the same time. The Salamander, however, is able to kill whole peoples." In other books one may read that the salamander is able to extinguish fire.

In Germany in the 17th century a wife tried to kill her husband by cooking a salamander in his soup. She was caught, seized and condemned to death, not for the attempted murder, but for using witchcraft.

It has only been during the past 80 years that animal venoms have been the subject of intense scientific investigation, at first from the clinical and pharmacological point of view, then from the chemical. Snake venoms first stood in the foreground of research, and the names of Vital Brazil and the "Instituto Butantan" in Sao Paulo as well as of Albert L. Calmette and the "Institut Pasteur" in Saigon are due special mention. Since the mid-twenties of this century, the chemistry of animal venoms has been studied intensively. The pioneering work in this area were investigations on toad venoms by Heinrich Wieland and his group in Munich and Freiburg. The book "Die Biochemie der tierischen Gifte" by E. Kaiser and H. Michel, which appeared in Vienna in 1958, greatly stimulated the field. A few years later, so many laboratories all over the world were investigating medical, biological and chemical aspects of these venoms that an international symposium on animal toxins and toxic animals was held in Sao Paulo (1966) and the International Society on Toxinology was founded (1964).

In the meantime it turned out that many more animals are venomous or poisonous than originally thought. Our conception of "toxins" had to be revised, too: we must not only consider their pharmacological activity, but also a defined and ecologically relevant function in biology, be it for catching prey or protecting against enemies. Consequently one must distinguish between "actively" and "passively" venomous animals. The latter may be divided into "primary venomous" animals that possess organs especially developed for defense (e. g. amphibia, beetles) and "secondary venomous" or "poisonous" animals that happen to be toxic during certain seasons of the year and usually become toxic by chance via the food-chain (e. g. shell-fish or certain fishes).

In this connection the question of "toxicity" arises. A measure for it is the LD_{50}; that is the lethal dose for 50% of test animals.

Mice are generally used for these tests, but the values found by this way must not be transferred to other animals or men. But at least they give an idea of the order of magnitude.

In this book medical considerations will receive extensive coverage first because the number of incidents involving poisonous animals is increasing (for various reasons), and second because modern clinical methods for treating attacks by venomous animals need to be reviewed. This is particularly important because many books deal with treatments that are more akin to mythology or folk medicine than modern science. The number of such has decreased in recent years, but even in recognized journals from time to time treatments are dealt with that are useless or even dangerous, especially if they delay effective therapy. How often do significant cases of poisoning occur from bites or stings?

It is frequently accepted that every sting or bite necessarily leads to a toxic reaction and that such a reaction is usually fatal. Fortunately this is wrong. According to reliable statistics and calculations, the average rate of mortality is around 2.5%. But even without fatal consequences, a bite or sting may cause permanent damage to the afflicted body part. Animal venoms exhibit a fascinating variety of pharmacological actions and chemical structures. In addition to the widespread peptides and proteins biogenic amines, alkaloids, heterocyclics, terpenes, steroids and glycosides – a broad spectrum of substance classes may be found. A corresponding breadth of pharmacological activity is seen: Besides numerous enzymatic activities, true toxic effects can be observed that are caused by cardiotoxins, neurotoxins, hemotoxins or cytotoxins. The study of the biosynthesis of such compounds puts evolutionary relationships in a new light and serves as an additional criterium for the systematic biologist. As interesting as these animals are for the scientist, they are potentially very dangerous for the inexperienced and careless laymen, especially travellers and vacationers. Among the species that are known to provide the unwary with a nasty experience are the venomous marine animals. Sea urchins are at least visible, and one may avoid them; more insidious are the sting rays, or stone fishes, which are quite often hidden in the sand; or doctor fishes or scorpion fishes, which defend their hunting ground and may attack divers. Quite malicious are the nettle cells of coelenterates i. e., jelly fish, sea wasps, sea anemones, sea nettles and corals, especially the tentacles of jelly fish, which may cause severe burns even after separation from the animal. About 50,000 incidents involving marine animals are estimated to occur worldwide each year. This figure does not cover the 20,000 cases of intoxication by consumption of poisonous fish or shellfish. Fortunately nearly all of these accidents are not severe; the number of fatal cases probably does not exceed 300. In many parts of the world snake bites are not uncommon occurrences. 1.7 million snake bites may occur per year, 40,000 resulting in fatality. About half of these cases (in which the circumstances were known) were caused by carelessness, and the saying "an ounce of prevention is worth a pound of cure" certainly rings true here, too. Such preventive measures merely apply common sense. Snakes should never be frightened; they are usually more afraid of humans than vice versa, and they most often bite in selfdefense. In case of any doubt go out of their way and restrict your contacts to reading about them in books such as these!

Protection from Snake Bites

An active immunization against snake toxins is practically impossible. Hence precaution is the best protection. Most snake bites occur on the feet (60%) as a result of walking barefoot. 77% of bites are on the shank (including the foot) and could have been avoided by wearing boots. Considering that more than 96% of all snake bites occur on feet, legs and hands, one may see what an important role carelessness plays or how important precaution is. Good precautionary measures are:

1. Not walking barefooted.
2. Stepping hard when walking.
3. Not turning or lifting stones with your hands, unless you are sure that no snake is hidden under them. Use a stick instead.
4. Not reaching into holes without previous inspection.
5. Being especially careful about snakes after sunset.
6. Not attempting to needlessly kill a snake; many snakes bite in self-defense.
7. Not swimming in waters where snakes abound. Many land species of snakes swim well and may bite while in water.

The number of fatal snake bites may seem high; however, the number of fatal bee and wasp stings is about the same. On the

American continents "snake bite kits" are available for first aid; they contain serum and syringes. Their manufacturers are listed on page 194, and they may be obtained at drugstores, from physicians or from the producer directly. It is usually necessary to indicate the geographic area of travel when ordering such a kit in order to get the special sera. A layman without experience in injecting sera should never try to utilize these kits.

Stings and bites from scorpions and spiders are so frequent that one can hardly obtain reliable figures on incidence. In Mexico, about 70,000 scorpion stings per year are estimated, 1200 of them fatal. Except for the *Uloboridae,* all spiders are venomous; few species, however, really pose a danger to humans. The fangs are generally not strong enough to penetrate human skin, but in any case one should avoid touching them with unprotected hands. Scorpions and spiders usually are hidden during daytime, sometimes under a thin layer of sand. Again, it is advisable not to walk

Table 1. Lethal bites and stings by venomous animals in the USA from 1950 until 1954

Animals	Number	Percent
Insects	86	40.0
Bees	52	24.2
Wasps	21	9.8
Yellow jackets	7	3.2
Hornets	5	2.3
Ants	1	0.5
Poisonous snakes	71	33.0
Rattle snakes	55	25.6
Mocassins (Cottonmouth)	2	0.9
Coral snakes	1	0.5
Unidentified	13	6.0
Spiders	39	18.1
Scorpions	5	2.3
Nettle animals	1	0.5
Sting rays	1	0.5
Yet undefined animals (insects, spiders, scorpions)	12	5.6
Total	215	100.0

barefoot. They come out to hunt in the evening, a time of many accidents, requiring extra precaution. Clothes, shoes or food placed on the floor are a special invitation to these animals. In dangerous areas such items should be carefully sealed in a plastic bag. A rather reliable and representative review of incidents involving venomous animals in the USA (1950–1954) is given in table 1.

The purpose of this book is not to convince the reader that danger from venomous animals lurks around every corner. Rather, it is to show how to avoid accidents and behave in case one occurs.

And not everything that looks venomous should be killed. Many of these animals have ecologically important functions and are even protected by law because their importance far outweighs the occasional harm they cause.

1 Coelenterates, Cnidaria

Distribution

Cnidaria and *Acnidaria* together form the phylum *Coelenterata.* Here we need to consider the cnidarians or nettles and the venomous polyps, jelly fish, sea wasps, sea nettles, sea anemones and corals. They are found in all seas within a range of about 45° North to 30° South with small geographic exceptions in the North Atlantic due to the Gulf Stream. Table 2 presents summaries on the most important species.

Poisoning

Poisonings are caused by nettle capsules (nematocysts), which are formed in special cells, the so-called cnidoblasts. The nettle capsules are localized primarily on the tentacles; they are also found, however, in the epidermis of the mouth region as well as in some inner structures. The structure of the nematocysts is remarkably complicated in relation to the size (usually less than 1 mm). When the cnidocils are touched the thread tube is ejected and injected into the skin of the victim; at the same time the venom pours out. It consists of several peptides of relatively low molecular weight that are especially toxic to crustaceans and fishes. Thus, for example, 1 ng/kg of the toxin of the sea anemone *Condylactis gigantea* or 2 ng/kg of the toxin from *Anemonia sulcata* causes paralysis of crabs and shrimps.

The symptoms of coelenterate poisoning vary according to species, site of sting and sensitivity of the victim. As numerous studies have shown, repeated stings result in an increased sensitivity that may finally give rise to an anaphylatic reaction.

Contact with the tentacles leads to symptoms that range from itching to an intense burning; the pain may become so intense that

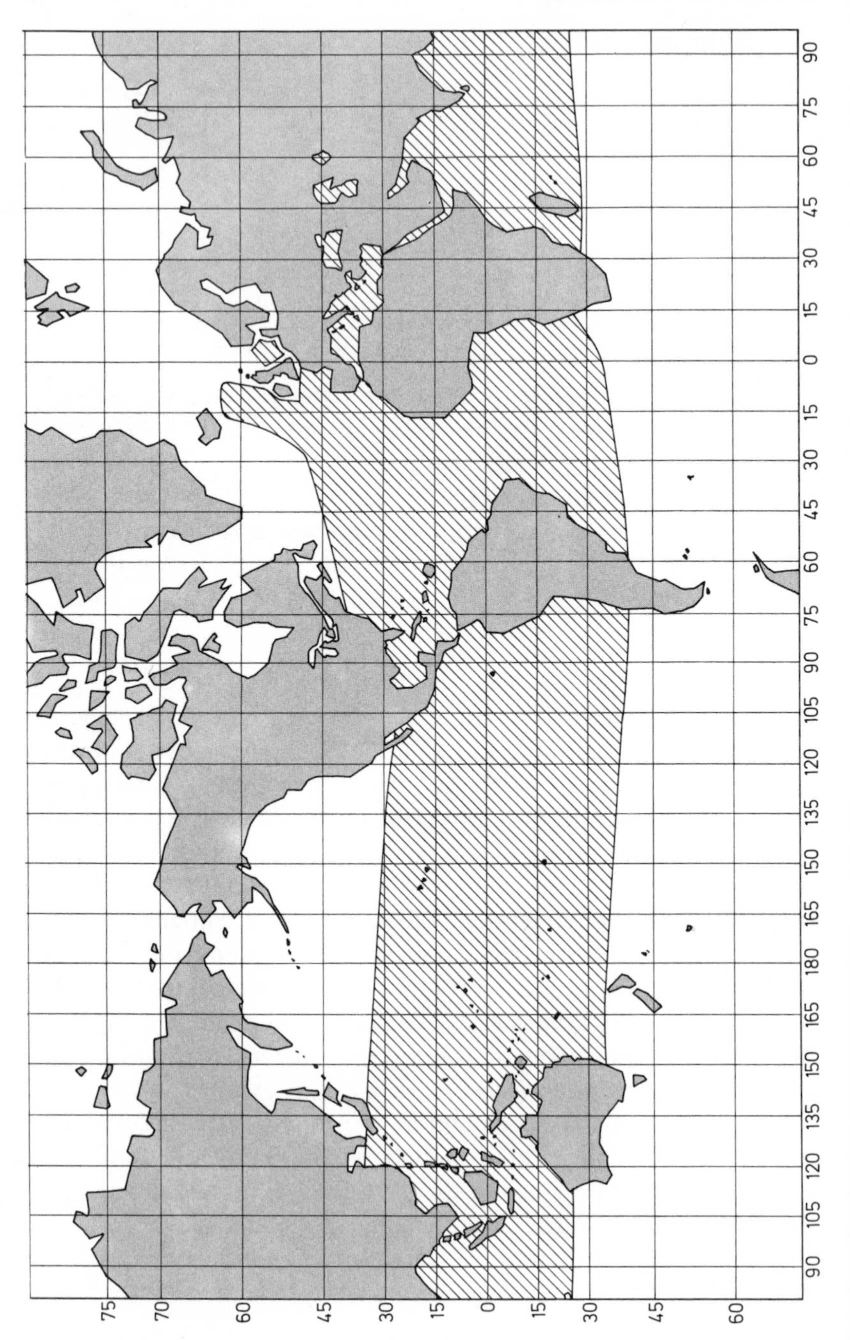

Fig. 1. Distribution of coelenterates

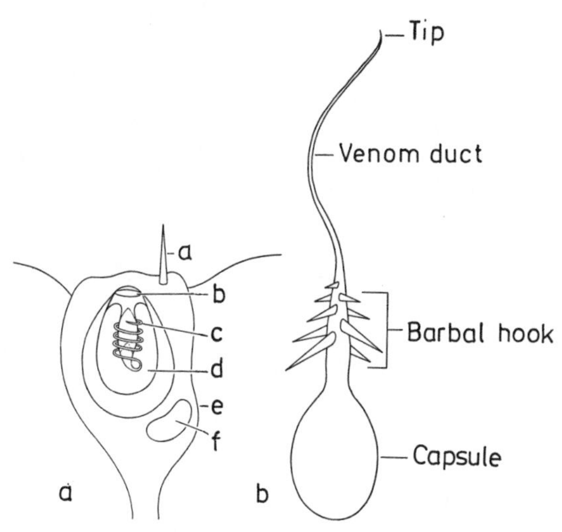

Fig. 2a, b. Nematocyst. **a** in rest, *a* cnidocil, *b* operculum, *c* venom duct, *d* venom capsule, *e* cnidoblast, *f* nucleus; **b** in the ejected state

victims fall unconscious. The initially local pains change later on to numbness or into hypersensitivity.

Sagartia stings at first cause nettle-like eruptions and later on necrotic ulcers. Stings of *Scyphozoa* may be harmless, as in the case of *Cyanea* species native to the North Sea and Baltic Sea, or painful, as in the case of other *Cyanea, Catostylus* and *Chrysaora* species. Here generalized symptoms are also observed: primary shock, collapse, headache, shivers and fever. In serious cases muscle seizures, paralysis and finally death from heart failure may occur. The same symptoms may be observed after *Physalia* stings.

Fatal stings in humans are rather rare, but when they occur death is usually very rapid, mostly within a few minutes. Victims surviving the first 30 minutes usually recover within a few hours to several weeks. Without doubt the most dangerous among those animals is the sea wasp, *Chironex fleckeri;* the venom is cardiotoxic, and the sting causes hypertension, lung edema and acute heart failure. In fact, the sea wasp probably is the most dangerous venomous marine animal of all. As death occurs within seconds or at most minutes after touching the tentacles, treatment is impossible. Divers must be extremely careful, especially since even adult animals are

Table 2

A. Hydrozoa	
Milleporidae:	*Millepora alcicornis L.* Distribution: Tropic Pacific, Arabian Sea, Caribbean Sea
	Millepora platiphylla Distribution: West Indian Islands
Physaliidae:	*Physalia physalis L.* (man-o-war) Distribution: Atlantic from the tropics to the Hebrides, Mediterranean Sea
	Physalia utriculus Distribution: Indo-Pacific, South Japan, Hawaii
B. Scyphozoa	
Carybdeidae:	*Carybea alata Reynaud* (sea wasp) Distribution: Tropic Pacific, Atlantic, Arabian Sea
	Carybdea manupialis L. Distribution: Atlantic, West Indies, West Africa, Portugal, Mediterranean Sea
Catostylidae:	*Catostylus mosaicus* Distribution: Australia, Philippines, New Guinea
Chirodropidae:	*Chironex fleckeri* (sea wasp) Distribution: Northern Australia, Arabian Sea, Philippines
	Chiropsalmus quadrigatus (sea wasp) Distribution: Northern Australia, Arabian Sea, Philippines
Cyaneidae:	*Cyanea capillata L.* (sea nettle) Distribution: North Atlantic, North Sea, Baltic Sea, Alaska, Japan, China
	Cyanea lamarcki Distribution: North Sea from France to Norway, North Atlantic, North Pacific
Pelagiidae:	*Chrysaora quinquecirrha* (sea nettle) Distribution: West Africa, Azores, Eastern Coast of USA, Arabian Sea, West Pacific
C. Anthozoa	
Actinidae:	*Actinia equina* (sea anemone) Distribution: East Atlantic, Mediterranean Sea, Black Sea
	Anemonia sulcata Distribution: East Atlantic, Mediterranean Sea, Black Sea
	Condylactis aurantiaca Distribution: Mediterranean Sea
Sagartiidae:	*Sagartia elegans* (sea anemone) Distribution: East Atlantic from Iceland to France, Mediterranean Sea

not more than eight inches in diameter; the body is light blue and opaque and therefore hardly visible in water. In addition the tentacles may be several feet long and remain dangerous even if torn off the body.

Treatment

First aid measures must aim at inactivating the tentacles and slime containing the nematocysts; ethanol, 10 per cent formaldehyde (aqueous solution), dilute aqueous ammonia or sodium bicarbonate can be applied. In emergencies in which these substances are not available, sugar, salt, olive oil or *dry* sand may be applied to the affected site. They must be allowed to dry before being scraped off with the back of a knife or a piece of wood. Fresh water and wet sand may not be used, because they stimulate further ejection of nematocysts.

In severe cases the victim should be placed on his back; artificial respiration and heart massage should follow. To retard the absorption of the toxin a constriction band or a tourniquet should be placed upon the affected joint. It must be applied tightly enough to occlude the superficial veins and lymphatic return but not arterial flow. A constriction band, however, should not be applied for more than four hours. Further measures consist of analgesic treatment, e. g., with morphine sulfate, treatment of neurotoxic effects and control of the primary shock. Injections of calcium gluconate compensate muscle seizures. Experiments to produce antivenin against *Chrysaora, Physalia* and *Chironex* seem so far to be promising. These antivenins act only on the general symptoms; neither skin necrosis nor the strong local pain can be cured. Poisonings from *Cyanea* spp. of the North Sea and the Baltic Sea usually are harmless; they can be treated locally with dilute aqueous ammonia.

Chemistry

The crude venom of *Chrysaora* is a viscous fluid, from which a lethal protein fraction has been isolated; the molecular weight is 150,000. It probably consists of eight sub-units of one smaller protein of molecular weight 19,000. In addition to this toxin, at least seven enzymes are found. The same is true for *Physalia:* one toxic protein of molecular weight 150,000 and several enzymes. On the other hand the active peptides of *Chironex* possess molecular weights between 10,000 and 30,000 only. The toxin of *Cyanea capillata* not only acts as a cardiotoxin but also possesses a dermatonecrotic effect. It is a polypeptide of molecular weight 70,000; the LD_{50} is 0.7 mg/kg (mouse, i. v.).

The most striking and clinically most important effect of the coelenterate toxins are the extremely strong local pains. In the case of *Chrysaora* they are due to histamine, serotonin, prostaglandines E and F, and peptides having kinin-like action. With *Physalia* the pains are caused by kinins exclusively; histamine, serotonin and prostaglandins have not been found. More complex are the pain producing factors of *Chironex,* among which histamine and kinins are again observed. The skin necrosis is produced by proteins of molecular weights between 50,000 and 300,000. Three toxins have been isolated from *Anemonia sulcata* in pure form; they show neurotoxic and cardiotoxic effects in mice. They are extremely toxic, however, against the crab *Carcinus maenas* with an LD_{100} of 2 ng/kg for toxin I and II, and 50 ng/kg for toxin III.

Toxin I contains 45 amino acid residues; it possesses a molecular weight of 4702. The amino acid sequence is:

Gly-Ala-Ala-Cys-Leu-Cys-Lys-Ser-Asp-Gly-Pro-Asn-Thr-Arg-Gly-Asn-Ser-Met-Ser-Gly-Thr-Ile-Trp-Val-Phe-Gly-Cys-Pro-Ser-Gly-Trp-Asn-Asn-Cys-Glu-Gly-Arg-Ala-Ile-Ile-Gly-Tyr-Cys-Cys-Lys-Gln

Toxin II with 47 amino acid residues and a molecular weight of 4770 is the main component and possesses the following sequence:

Gly-Val-Pro-Cys-Leu-Cys-Asp-Ser-Asp-Gly-Pro-Ser-Val-Arg-Gly-Asn-Thr-Leu-Ser-Gly-Ile-Ile-Trp-Leu-Ala-Gly-Cys-Pro-Ser-Gly-Trp-His-Asn-Cys-Lys-Lys-His-Gly-Pro-Thr-Ile-Gly-Trp-Cys-Cys-Lys-Glu

Toxin III with 24 amino acid residues (m. w. 2678) has the following sequence:

Arg-Ser-Cys-Cys-Pro-Cys- Tyr-Trp-Gly-Gly-Cys-Pro-Trp-Gly-Glu-Asn-Cys-Tyr-Pro-Glu-Gly-Cys-Ser-Gly-Pro-Lys-Val

References

Halstead, B. W.: Poisonous and Venomous Marine Animals Vol. 1, U. S. Govt. Printing Office, Washington, D. C., 1965

Schneider, W. P.: Prostaglandins from Marine Sources. Food-Drugs from the Sea Symposium 1972, p. 151–155; Proceedings edited by L. R. Worthen

Toom, P. M. et al.: Toxicon *13,* 158 (1975)

Lane, C. E.: A. Rev. Pharmac. *8,* 409 (1968)
Randt, C. T. et al.: Science *172,* 495 (1971)
Beress, L., Beress, R.: Kieler Meeresforschungen *27,* 117 (1971)
Beress, L., Beress, R., Wunderer, G.: Toxicon *13,* 359 (1975)
Wunderer, G., Machleidt, W., Wachter, E.: Hoppe-Seylers Z. physiol. Chem *357,* 239 (1976)
Burnett, J. W., Calton, G. J.: Toxicon *15,* 177 (1977)
Rathmayer, W., Jessen, B.: Naturwissenschaften *62,* 538 (1975)
Rathmayer, W., Beress, L.: J. comp. Physiol. *109,* 373 (1976)
Walker, M. J. A.: Toxicon *15,* 3, 15 (1977)

2 Mollusca (Molluscs)

2.1 Lamellibranchiata (Mussels)

Shellfish poisonings are among the most frequent and widespread of toxic reactions. They are so striking and evident that they were already recognized as a medical problem in the 17th century. Since then numerous cases have been reported; they occur especially in North America and Europe, but also in East Asia. The belief that such poisonings are caused by spoiled shellfish is usually not true. Well defined toxins are far more often responsible, toxins either produced by the shellfish itself or absorbed from the food and stored for some time. It has been well known for a long time that the toxicity of shellfish runs parallel to the appearance of dinoflagellates; this explains the seasonality of the toxicity of shellfish that are otherwise edible. The months from May to September are most dangerous, while during the rest of the year practically no cases of poisoning are observed. The popular belief that one should not eat shellfish in the months "without R" is probably derived from this observation. Contrary to widespread claim, it is not possible to discriminate between poisonous and non-poisonous shellfish by odor or discoloration of a silver spoon. Also, the usual methods of preparation do not destroy the toxins. A certain detoxification can be observed by boiling in water after addition of a tablespoon full of sodium bicarbonate per quart of water. This treatment, however, causes some of the typical taste of shellfish to be lost. It is also used in canneries, in addition to toxicologic tests, as a safety measure, so that canned shellfish may be eaten without hesitation.

The following species are most frequently involved in poisoning:

Saxidomus giganteus (Butter clam) western coast of USA from Alaska to California

Saxidomus nuttalli	(Sand cockle) coast of California
Ostrea edulis	(oyster) European coasts
Mytilus edulis	(sea mussel) European coasts; eastern and western coast of USA
Mytilus californianus	(Californian mussel) North Pacific up to the Aleutian

Poisoning

According to symptoms, three kinds of poisoning may be observed:

1. Gastrointestinal poisoning. This is characterized by a relatively long latency period of about 10–12 hours. Symptoms consist of nausea, vomiting, diarrhea and abdominal pains. Recovery is usually rapid, and the poisoning is without any consequences.
2. Erythematous poisoning. The latency period is short, usually 2–3 hours. Symptoms are connected with allergic reactions: diffuse erythema, swelling and urticaria that particularily affects the face and neck, but may subsequently involve the entire body. Headache, sensation of warmth, conjunctivitis, dryness of the throat, swelling of the tongue and dyspnea may be present, too. The patients usually recover a few days later; lethal cases are extremely rare.
3. Paralytic shellfish poisoning. This type of poisoning, in contrast to the preceding ones, is decidedly serious; about 8% of the cases are fatal. The responsible agent is saxitoxin, a neurotoxin produced by the dinoflagellate *Gonyaulax catenella* and *G. excavata,* which is taken up by the shellfish from nutrients. Symptoms usually appear within 30 minutes. Initially there is a tingling and burning sensation in the face, especially of the lips and tongue, which proceeds from the neck to the fingertips, or over the whole body down to the feet. This paresthesia later changes to numbness.

In severe cases ataxia and general motor incoordination are accompanied by a peculiar feeling of lightness, "as though one were floating in air." Weakness, confusion, headache, rapid pulse, intense thirst and myalgia may occur as well, whereas gastrointestinal symptoms (see heading 1) are less common. In the severest cases dysopia and even temporary blindness occur. Death occurs as a result of respiratory paralysis, usually within a period of 12 hours. Beyond 12 hours the chance of surviving is good. The lethal dose for man is about 1 mg (calculated for a body weight of 75 kg).

Toxins other than saxitoxin were observed in *Gymnodinium breve, Amphidinium carteri* as well as in the two species mentioned above. However, they have not yet been investigated. Most dinoflagellates tested so far were nontoxic. This topic deserves careful study in future.

Treatment

Since there is no specific antidote, treatment has to be restricted to the symptoms. Saxitoxin and other shellfish toxins are readily absorbed on activated charcoal, hence this can be recommended. The use of drugs is controversial, but neostigmine, ephedrine and DL-amphetamine have been shown to be useful; digitalis preparations, however, are not advisable. Artificial respiration may be necessary. Under no circumstances should alcohol be consumed during or some time after the poisoning.

Chemistry

The structure of saxitoxin is known; it is a low molecular weight compound, resembling to some extent e. g. the guanidine residues of tetrodotoxin (p. 103):

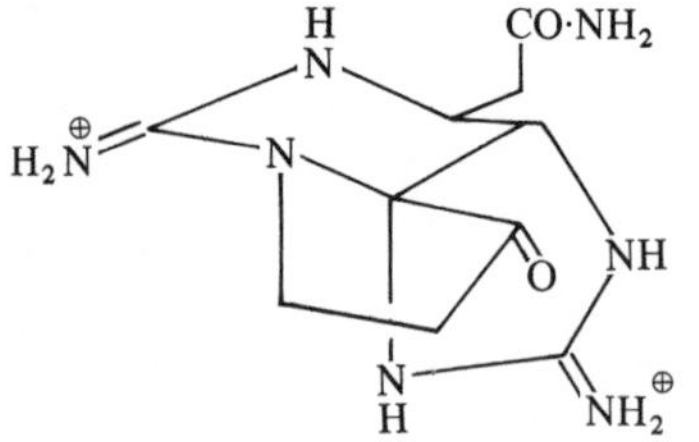

Saxitoxin

2.2 Gastropoda (Snails, Slugs)

2.2.1 Snails

A toxin causing symptoms similar to those of saxitoxin has been isolated from the marine snail *Babylonia japonica,* which is used as food in Japan. It was found that after living for a short time in

Suruga Bay, nonpoisonous animals became inedible; on the other hand poisonous animals from Suruga Bay lost their toxicity after living in another area or in clean water for a few months. Hence in this case, too, algae or other small organisms in the food chain must be suspected as contributors to the origin of the toxin. Fortunately, fatal cases have not yet been recorded.

Symptoms and Treatment

Symptoms of intoxication may include: visual impairments, including amblyopia and mydriasis, thirst, numbness of lips, speech disorders, constipation and dysuria.

Treatment is usually symptomatic; the patients recover within a few days.

Chemistry

The structure of the toxin has been elucidated. It is a low molecular weight compound of unique structure without any analogy.

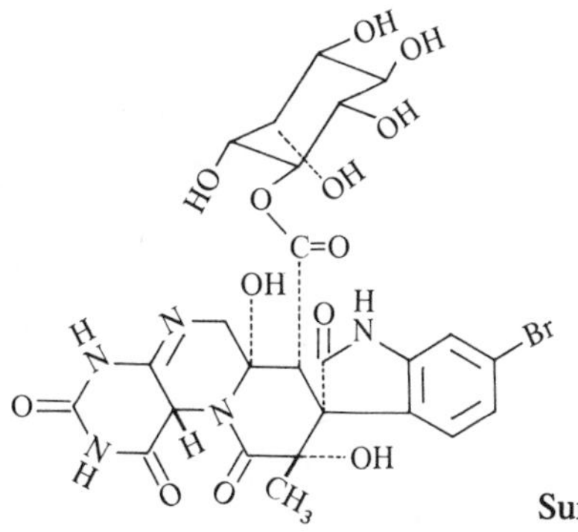

Surugatoxin

2.2.2 Toxoglossa, Conidae, Coneshells

Whereas snails and mussels are poisonous "by accident" and thus belong to the "passively venomous" animals, the toxoglossa possess a complicated and well-developed venom apparatus used for hunting prey; thus, they are an example of so-called "actively venomous" animals. Toxoglossa are divided in three families: *Conidae,*

Turridae and *Terebridae*, but only *Conidae* of tropical or neotropical seas are involved in cases of human poisonings. Those coneshells that feed upon fish are extremely dangerous. Of the approximately 500 species, the following have to be considered:

Conus achatinus	(East Africa)
C. aulicus	(Mauritius, Seychelles, Polynesia, Australia)
C. geographus	(Red Sea, East Africa, Polynesia Australia)
C. gloria-maris	(Philippines, East Indonesia, Fiji-Isles, Solomons, New Guinea)
C. magus	(East Africa, South Japan, Australia)
C. marmoreus	(East Africa, Polynesia, Hawaii, Ryukyu-Isles, Australia)
C. omaria	(East Africa, Red Sea, Hawaii up to the Midway-Isles)
C. striatus	(East Africa, Red Sea, Hawaii, Polynesia, Australia)
C. textile	(East Africa, Red Sea, Hawaii, Polynesia, Australia)
C. tulipa	(East Africa, Polynesia, Australia)

Most accidents involve *C. geographus* and *C. tulipa*. The only European species, *C. mediterraneus,* is completely harmless. It is native to the Mediterranean Sea, Algarve Coast, Canary and Cape Verde Islands, and the coast of West Africa.

Poisoning

Cases of poisoning from coneshells are not a major public health problem considering the figures; only about 30 have been reported so far, eight of them fatal. Most of the accidents happen as a result of carelessness when collecting the beautiful shells. Leather gloves should be worn for collecting as well as for removing the body. Coneshells sting with tiny harpoon-like teeth (Fig. 3) produced in a special sack, where they are stored until they are needed. One of these dart-like structures is then moved through the pharynx to the end of the proboscis; on the way it is filled with the venom when passing the venom gland. It is finally projected from the proboscis into the body of the victim.

The stings produce a puncture-like wound that is extremely

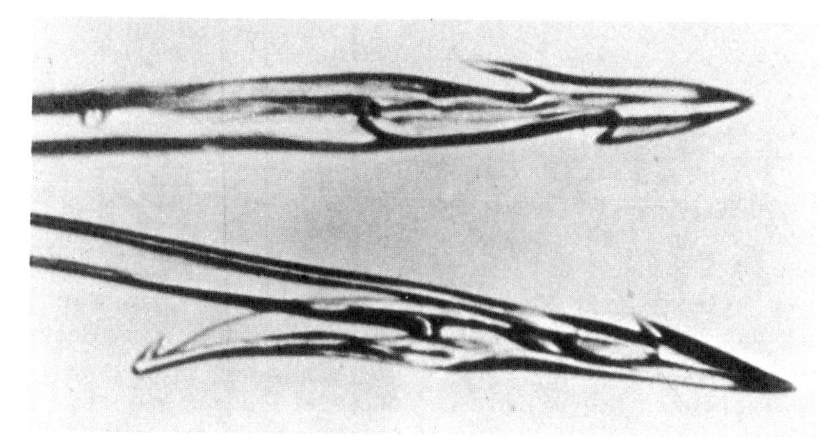

Fig. 3. Tip of the harpoon-like sting of *Conus striatus*. The actual length is about 7 mm, the diameter 0,5 mm

painful; the site of the sting swells, and the pains change gradually into paresthesia that spreads over the whole body, especially the lips and mouth. In severe cases a muscle paralysis appears as the next symptom. Visual disturbance and vomiting have also been observed; the cause of death is heart failure.

Treatment

No specific antidote is available, and the poisoning can only be treated symptomatically. Artificial respiration may be necessary. The recovery period varies between one day and several weeks; the patients, however, complain for a long time of general weakness and rapid fatigue from physical exertion.

Chemistry

Toxins of *C. magus* and *C. achatinus* have been under study for the past few years. As far as we know, they are heat labile myotoxins of a molecular weight of about 10,000. Detailed investigations have not as yet succeeded because of lack of sufficient pure toxin.

2.3 Cephalopoda (Cuttlefishes, Squids and Octopuses)

2.3.1 Octopoda

The octopus *Hapalochlaena maculata* (= *Octopus maculosus)* is found on the Australian east coast. The bite of this "blue-ringed octopus" is painless, and thus usually not noticed. There are, however, severe systemic symptoms such as numbness of limbs, vomiting, temporary blindness and loss of muscle coordination; death may occur by respiratory paralysis.

The venom responsible is produced in the anterior salivary glands; it is used for paralyzing or killing of prey, mostly small crabs.

The toxic substance has been isolated and was named maculotoxin. Structural analysis has shown that it is identical with tetrodotoxin.

References

Freeman, S. E., Turner, R. J.: Br. J. Pharmac. *46,* 329 (1972)
Turner, R. J., Freeman, S. E.: Toxicon *12,* 49 (1974)
Halstead, B. W.: Poisonous and Venomous Marine Animals, Vol. 1, U.S. Govt. Printing Office, Washington, D. C., 1965
Kosuge, T. et al.: Tetrahedron Letters, 2545 (1972)
Simon, B., Mebs, D., Gemmer, H., Stille, W.: Dtsch. med. Wschr. *102,* 1114 (1977)
Bates, H. A., Hostriken, R., Rapoport, H., Toxicon *16,* 595 (1978)

3 Arthropoda

3.1 Arachnidae

3.1.1 Scorpiones (Scorpions)

Taxonomy of scorpions

Order: Scorpiones

Suborder: Buthoids	and	Chactoids
Family Buthidae		Chactidae
Buthinae, Tityinae *(Subfamilies)*		Scorpionidae
		Diplocentridae
		Bothriuridae
		Vegovidae

The scorpions are a class of the animal kingdom containing about 650 species. The smallest of them are about 1 inch, the biggest up to 10 inches in length. They inhabit large parts of the southern United States, Mexico, Central America including the Caribbean Islands, the Northwest of South America and large parts of Brazil. In the Old World they are native to the whole Mediterranean area, North, West and East Africa, South Africa, Madagascar and Arabia (Figs. 4 and 5).

Scorpions had an important status in the ancient mythology of the Old World. They are mentioned in the Greek mythology, (Orion legend), in the Talmud and in the Bible; they also played an important role in the Mithras cult.

The symptoms of poisoning vary greatly, which according to species, may be completely harmless or on the other hand even fatal. Scorpions, like spiders and other venomous animals, form

Fig. 4. Geographic distribution of dangerous scorpions in America (according from Bücherl 1971)

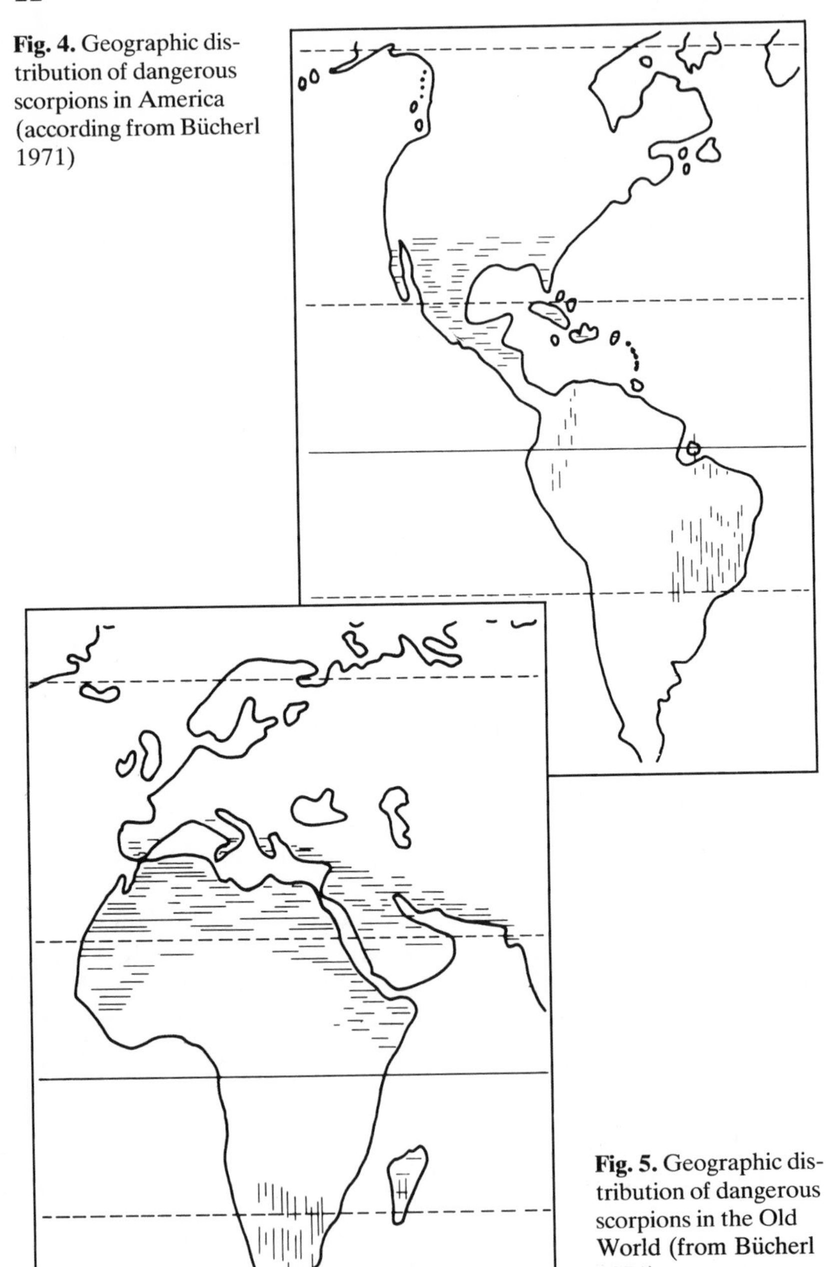

Fig. 5. Geographic distribution of dangerous scorpions in the Old World (from Bücherl 1971)

geographic races, which differ remarkably in the activity of their venoms. For this reason the place of origin of these animals is a very important consideration in any investigation.

Buthoids

With few exceptions scorpions dangerous to humans belong to the genus *Tityus* (North and South America), *Centruroides* and *Centrurus* (Mexico), *Androctonus, Buthacus, Leiurus, Buthotus* and *Buthus* (Mediterranean area and North Africa) and *Parabuthus* (South Africa). They sometimes occur in very large numbers, even close to settlements, so to speak as unwanted domesticated animals. Reliable estimates come to about 150,000 accidents per year (world-wide); 70,000 in Mexico alone, 1,200 of these fatal. In Brazil the death rate for adults in 0.8–1.4%, children 8 to 16 years old 3–5%, and small children 15–20%. In the city of Belo Horizonte, Brazil, 874 cases were recorded in one year with 100 fatalities.

The species native to the North coast of the Mediterranean Sea (France, Italy, Yugoslavia) are more or less harmless. However, in West Africa five dangerous species are found: *Androctonus australis, Buthus occitanus, Buthacus arenicola, A. amoreuxi* and *A. oenas. B. occitanus* also occurs in South France; it is an example of the observation mentioned above that the same species may differ in toxicity from place to place. Whereas *B. occitanus* is very dangerous in North Africa, it is completely non-toxic in South France. 80% of all accidents and 95% of the fatal cases in North Africa are

Table 3. Breakdown of scorpion stings in Algeria, 1942–1958

Patients	Number of accidents	Number of lethal cases		% lethality
pre-school children	411	32		7.78
School children	5,211	191		3.66
Adults	13,130	145		1.10
Elderly	1,412	18		1.27
Total	29,164	386	average:	1.90

attributed to *A. australis.* A summary of scorpion sting incidence in Algeria is given in Table 3.

Accidents involving scorpions are generally a result of carelessness by the victim; they can happen indoors, in garden and fields, and in the desert. Since scorpions are nocturnal animals, special care must be taken after sunset. They prefer to crawl into clothing or shoes during the night; hence such items should be carefully shaken before one dresses in the morning. Scorpions usually dig themselves into sand during the daytime, but only under a thin layer. Walking barefoot thus is one of the most reliable ways to be stung by a scorpion. Good footwear is indispensable in the infested areas. Scorpions are most aggressive during hot and stormy weather.

Poisoning

A survey of the toxicity of some scorpion venoms is given in Table 4. Note that those values have been measured in mice; according to the independent observations of several authors, humans are much more sensitive.

The sting of a scorpion usually causes severe pain at the sting site. The victim becomes excited, and in children convulsions may be observed; many reflexes become diminished or vanish. Consciousness is usually retained; feelings of fright and fear are frequent. Eyes run with tears; visual power is restricted. The pulse rate is increased and irregular, the blood pressure may be higher or lower, the body temperature is instable; lasting hypothermia is a sign of aggravation of the general condition. Breathing becomes irregular. Vomiting is a serious sign that the nervous centers have been affected. Death occurs from respiratory paralysis, usually within the first 20 hours, rarely after up to 30 hours. It is very important that the patient be monitored for at least 12 hours after the last symptoms have disappeared; in many cases after an improvement of the general condition a serious relapse may occur.

Centruroides spp.: In North America, Central America and the Western part of South America scorpions of medical importance belong to the genus *Centruroides* (synonym: *Centrurus). Centruroides* also occur on the Cape Verde Islands, Sierra Leone and Gambia. Scorpions of these species frequently hide under the loose bark of trees, in the crevices of stubs, in storage rooms, piles of brick

Table 4. Toxicity of some scorpion venoms

Species	LD_{50} in mg/kg (mouse, s.c)
Tityus serrulatus	1.45
T. bahiensis	9.35
T. trinitatis	2.00
Centruroides limpidus	5.00
Leiurus quiquestriatus	0.33
Buthacus arenicola	3.5
B. leptochelis	0.77
Androctonus australis	6.00
A. oeneas oeneas	0.31
A. mauretanicus mauretanicus	0.32
A. crassicauda	0.40
A. amoreuxi	0.75
Buthus occitanus tunetanus	0.99
B. occitanus paris	4.15
Buthotus judaicus	8.00
Buthotus minax	4.25
Parabuthus spp.	35–100
Parabuthus transvaalicus	4.25
Opistophthalmus spp.	600.0
Hadogenes spp.	1800.0
Scorpio maurus	inactive (> 141.6)

and lumber or under rocks. They also frequently invade human dwellings. They crawl on walls and ceilings in search of prey, and as they are mostly nocturnal, they may fall on beds or hide in shoes or clothes. Their nocturnal life can be studied with the aid of ultraviolet light, since they are fluorescent.

In past decades scorpion stings were the most frequent type of accidents involving venomous animals: They were responsible for twice as many deaths as all those caused by snakes, spiders and insects. As a result of educating the public (eg, through television), tight house construction and other means of protection, scorpion stings are less frequent; and because of improved therapeutic practice, a decrease in mortality can be observed.

The venom of *Centruroides* does not produce visible effects at the site of the sting, but there is intense local pain. General symptoms are nervousness, respiratory difficulties and spasms. Excessive salivation, gastric hyperdistention accompanies difficulty in swallowing and speaking. Visual impairment includes mydriasis, diplopia and temporary blurred vision or even temporary blindness. In severe cases hypertension, cardiac failure and respiratory arrest may occur. The severity of the symptoms depends on age and constitution of the victim.

Therapy should consist only of antivenin. Incision and suction is of no value and may be harmful. The use of narcotics or analgesics is as strongly contraindicated as the drinking of alcohol. Barbiturates, however, have been used with good success.

Buthinae: The Buthid scorpions of the Old World belong to the *Buthinae,* a subfamily of the large family of *Buthidae* with 42 genera and more than 500 species. Distribution of the different species often depends on biotopes and climate. Thus, all species of *Buthacus* prefer sandy soils, whereas *Androctonus, Buthotus* and *Compsobuthus* prefer stony desert, loesslike soil or terra rossa. *Buthotus* and *Compsobuthus* are found under the bark of trees, *Buthus* and *Leiurus* are frequently found indoors. The scorpions are very resistant to their own venom; apparently the hemolymph of the scorpion is able to neutralize the venom. As a last resort, if specific antivenin is not available, hemolymph may be used if the condition of the victim is desperate.

The most important scorpions of the subfamily *Buthinae* are listed in Table 5.

The symptoms of envenomation from these scorpions are more or less similar. Cardiovascular disturbances are observed, such as peripheral vascular collapse, congestive heart failure, pulmonary edema, respiratory spasms and indirect inotropic effects. These symptoms result from the action of the toxin on the sympathetic

Table 5. Distribution of venomous scorpions, Buthinae

Name	Distribution
Buthus spp.	India, Algeria, North Africa
Palamneus spp.	India
Prionorus spp.	Algeria
Leiurus quinquestriatus	Israel, North Africa
Androctonus australis	North Africa

ganglia. Other clinical manifestations from direct action on the heart are bradycardia or tachycardia and arrhythmia leading to fibrillation. Blood pressure rises quickly, reaching values of 260/190 at a pulse rate of 150. Hyperthermia and thirst, urinary retention, mydriasis, disorder of speech and generalized erythema may also be seen. The sting is always connected with severe pain at the sting site.

Potent antivenin against *B. occitanus* and *L. quinquestriatus* are available; they are highly specific. Arguments concerning the usefulness of antivenin applications arose as a result of cases in which obviously an inadequate quantity of serum was used. Antiserum has to be injected intravenously in quantities of 15–30 ml., and even more (even in children). Agents blocking α- and β-receptors also have to be used according to the symptoms present.

Tityinae: Like the *Buthinae,* the *Tityinae* are a subfamily of the family *Buthidae.* The genus *Tityus* includes 63 species with 21 subspecies, living in South, Central and the Southern part of North America. They are practically in all parts of America, but in Brazil they form a public health problem. Thus, in one year (1935) in Belo Horizonte 2,449 cases were recorded with 145 fatalities. In the city of Ribeirao Preto, from 1945 to 1950 985 cases were registered. 64 of them were severe, and in those treated with antivenin only 7 deaths (small children) resulted. The statistics of Dr. Rosenfeld of the Hospital Vital Brazil at the Instituo Butantan, Sao Paulo, indicate that from 1954 to 1965, 1277 patients were stung by scorpions with only two fatalities; 701 of these cases were caused by *T. bahiensis, 36 by T. serrulatus.* In Trinidad, from 1929 to 1933 698 accidents were reported. The mortality rate was 0.8–1.4% for adults, 3–5% with school children and 15–20% among babies and small children. The most venomous species are:

T. serrulatus serrulatus
T. bahiensis bahiensis
T. trinitatis
T. trivittatus dorsomaculatus

They can be found under stones, in logs and loose bark of trees, in grapes, cellars, store rooms, in gardens and along rivers. Frequently they invade entire sections of cities, especially those with older houses.

Table 6. LD_{50} values of dry venom from *Tityinae* scorpions on white rats (wt = 20 g)

Species	Intravenously	Subcutaneously
T. serrulatus	0.016 mg	0.022 mg (Bücherl, 1953)
T. bahiensis	0.022 mg	0.045 mg to 0.140 mg
T. costatus	0.200 mg	1.100 mg
T.t. dorsomaculatus	0.014 mg	0.059 mg

Tityus stings immediately cause a transient burning sensation and intense local pain at the site of the sting as well as other pain in the entire afflicted limb that may last from some minutes to several hours. If the scorpion has not injected too much venom, which is usually the case, these local symptoms disappear within 24 hours. In more serious cases, however, systemic effects can be observed, such as progressive numbness, tightness in the throat, difficulty in speaking and involuntary twitching of the muscles. In severe cases convulsions and respiratory difficulties are observed; death is caused by respiratory paralysis.

All severe cases of envenomation must be treated with antivenom as soon as possible. A delay of two hours or more may decrease the success of this treatment. Again it is important to apply sufficient amounts of antivenom, ie, two to five ampoules intravenously and the same quantity subcutaneously.

Chemistry

Much progress has been made in recent years in the field of separation and structure determination of toxins of *Buthinae: Androctonus australis Hector, A. mauretanicus, Buthus occitanus paris, B. occ. tuneatus, Leiurus quinquestriatus quinquestriatus, Centrurus sculpturatus, C. suffusus.* 22 toxins were isolated and purified, and the structure of 19 of them determined, at least partially.

The venoms of *T. serrulatus* as well as *T. bahiensis* have been investigated chemically. The main problem was the separation and purification of the individual toxins, which was quite recently achieved by means of gel permeation in Sephadex. The principal toxin of *T. serrulatus,* according to initial findings, appears to be a polypeptide of 61 amino acid residues with a molecular weight of 6,996. The amino acid composition is:

Lys_7, His_1, Arg_1, Cys_6, Asp_9, Thn_2, Ser_3, Gly_4, Pro_3, Gly_5, Ala_4, Val_2, Ile_2, Leu_3, Tyr_6, Phe_1, Trp_2.

Though the toxins of African and Brazilian scorpions are different, they have several features in common; for example the absence of methionine, low histidine and phenylalanine content and a predominance of hydrophobic and basic amino acids.

Treatment

All serious cases, that is, stings of the species mentioned above, must be treated with antivenin. It is essential that either specific or polyvalent serum be given as soon as possible. A delay of just two hours may jeopardize success. The dose of serum must be sufficient to neutralize at least *2* mg of dry scorpion venom; in other words 5 to 10 phials. Half the dose should be given intravenously, the other half subcutaneously or intramuscular. Other treatments (tourniquet, sucking the venom out or cryotherapy) may be used as temporary aid only if the antivenom is not immediately available. Analgesics and symptomatic treatment are useful in support of serum therapy. All other measures must be avoided, as they are not only useless but in most cases cause additional damage.

Chactoids

Though the five families of Chactoids comprise about 60% of all species of scorpions, biochemical and physiological studies on them are far fewer than on Buthoids. Practically nothing is known about the venoms of Chactidae and Bothriuridae; they are harmless to man.

The behavior of chactoid scorpions is quite similar to that of the other scorpions. They are distributed in North America, India and around the Mediterranean Sea. Accidents often occur when scorpions have hidden themselves in clothes or shoes or when they are disturbed by children at play. Walking barefoot is always hazardous, and more than 70% of stings are localized on feet or hands.

Envenomations

Chactoid venoms are generally less toxic than those of *Buthinae* or *Tityinae*. On the other hand some genera like *Pandinus, Heterometrus* (Scorpionidae) and *Nebo* (Diplocentridae) are among the largest scorpions; they are able to inject considerable quantities of venom, and their stings are dangerous. Thus, for example, out of 75 cases of *Heterometrus* stings in India over 14 years, 23 fatalities were observed, among them nine adults. Other species dangerous to humans are *Scorpio* and *Urodacus* (Scorpionidae), *Hadrurus* (Vegovidae) and *Bothriurus* (Bothriuridae). Other scorpions usually considered harmless may occasionally provoke serious symptoms; these are *Euscorpius flavicandis* and *E. italicus* as well as *Vegovis confusus* and *V. spinigerus.*

In all cases the local symptoms are identical: severe pain and burning sensation, which may be followed by local paresthesia. In addition in the mild cases nose itching, salivation and dyspnea may occur. More serious symptoms are local edema, hemorrhages and muscle paralysis in the afflicted limb. Among the general signs, cardiovascular symptoms are prevalent; pulmonary edema may aggravate the case. The sting of *Euscorpius* may produce profuse perspiration, nausea, vomiting and hematuria.

Treatment

Because of the rarity of serious accidents there is no specific antivenin available. The symptoms mentioned above generally disappear within 24 hours. Many drugs have been recommended for soothing the pain and the swelling. Most of them have side effects. Highly recommended, however, is the injection of procaine at the site of the sting to relieve pain. In any case, rest, reassurance and maintenance of body warmth are important. "L. C." – Treatment (ligature and cryotherapy) may be applied. No other therapeutic measures, such as incision or suction, have appeared to be of value.

Chemistry

As already mentioned the venoms of chactoids have not yet been thoroughly investigated. The crude venoms consist mainly of three different types of compounds: the toxins, enzymes and biogenic amines.

Starting with the last group, 5-hydroxy-tryptamine (serotonin), 5-hydroxy-tryptophan, tryptophan und tryptamine and also histamine have been isolated from a number of venoms. Without any doubt, the serotin fraction is responsible for the elicitation of the pain.

The toxins are protein fractions and have been separated by electrophoresis and chromatography. The structure of none of them is clear. *Scorpio maurus, Heterometrus scaber, H. fulvipes, Nebo hierichonticus, Vegovis spinigerus, Hadrurus hirsutus, Pandinus exitialis, Hadrurus arizonensis* have been investigated.

With respect to enzyme activities, the hemolytic action of the venom of *Scorpionides* is well known. It is very strong in the venoms of *Hetreometrus scaber* and *Scorpio maurus,* weak in *Hadogenes spp.* Phospholipase A activity is also present in *Heterometrus scaber.* In addition acid phophatase, ribonuclease, 5'-nucleotidase, hyaluronidase and acetylcholinesterase were found, but no DNase nor alkaline phosphatase. Acetylcholinesterase has been found in *Vegovis spinigerus* and *Hadrurus mizonensis.* Proteolytic activities are variable.

References

Diniz, C. R.: Venomous Animals and their Venoms, Vol. III, chapter 54 (W. Bücherl and E. E. Buckley, Ed.) New York: Academic Press 1971

Bücherl, W.: ibid., chapter 55

Balozet, L.: ibid., chapter 56

Cheymol, J., Bourillet, F., Rock-Arveiller, M., Heckle, J.: Toxicon *12,* 241 (1974)

Hamilton, P. J., Ogston, D., Douglas, A. S.: Toxicon *12,* 291 (1974)

Ghazal, A., Ismail, U., Abdel-Rahman, A. A., El-Asmar, M. F.: Toxicon *13,* 253 (1975)

Mohamed, A. H., Darwish, M. A., Hani-Ayobe, M.: Toxicon *13,* 67 (1975)

Bücherl, W.: In: Arthropod Venoms (Ed. S. Bettini) p. 377, Springer-Verlag, Berlin Heidelberg New York, 1978

3.1.2 Araneae (Spiders)

Most of the 25,000 species of spiders are actively venomous; their venom apparatus consists of a pair of venom glands, venom ducts and claws with the aid of which the venom can be injected into the body of another animal. The venom is used for hunting prey but also for defense. In most cases by far, the activity and quantity of venom are too low to cause serious complications in humans.

On the contrary spiders are rather useful animals, and the antipathy towards them can hardly be justified, especially since they consume many noxious insects not only in the field, but also indoors. They are timid animals; they would rather flee humans than attack them.

There are, however, a number of spiders that present a real danger, and in some countries spiders are almost as much a public health problem as venomous snakes. This is true, for example, for Brazil, the Eastern Mediterranean area and Yugoslavia. Fatal bites in humans have been reported from *Trechona, Atrax, Harpactirella, Loxosceles, Latrodectus, Phoneutria,* more seldom *Mastophora, Chiracanthium* and *Lithyphantes.* Big and deep wounds rise from the bites of *Loxosceles, Acanthoscurria, Megaphobema, Xenesthis, Teraphosa, Avicularia, Phormictopus* and *Pamphobeteus.* Fewer consequences result from the bites of *Lycosa, Araneus, Argiope, Nephila, Tegenaria, Dendryphantes, Lasiodora, Grammostola.*

The number of accidents each year may amount to many thousands, several hundreds of them fatal. In this compilation one has to consider that by no means all species of these genera are dangerous and, as with snake bites, not every bite will take the worst turn. The consequences of a bite depend very much on the physical condition of the victim. The sex of the animal is also crucial: Severe poisonings are caused mostly by female spiders; males with the exception of Atrax, are not able, to deliver a deadly amount of venom.

The widespread fear of "tarantulas" (bird spiders) is completely groundless. They belong to the *Orthognata* and may be as big as 10 inches in diameter *(Acanthoscurria spp.).* They seldom bite; if so usually in defense. The bite itself often is no worse than a hornet sting, although the wounds are bigger, in proportion to the bigger venom chelizeres. But they present a different danger: These animals are not very clean and often feed on carrion; thus the chance of secondary infections is very high, which should be kept in mind in therapy. The dangerous tarantulas are *Harpactirella, Trechona* and especially *Atrax spp.* The latter, in particular *A. robustus,* are involved in many incidents.

They usually spend most of their lives in their funnel-like webs; during the mating season, however, they move about and even enter houses. If they become disturbed or frightened they assume a threatening, aggressive posture, and it is not uncommon for them to jump on their victims. The *Lasiodora spp.,* too, belong to the aggressive though not dangerous species. *Grammostola* on the other hand is peaceful and harmless. *G. mollicoma* is the largest known tarantula; its body may be 4 inches in length, the diameter including the legs more than 10 inches.

The distribution of venomous spiders is shown in Table 7 and Fig. 6.

Table 7. Distribution of venomous spiders

Genus or species	Country
Atrax	Australia, New Zealand
Harpactirella	South Africa
Loxosceles	South-, Central-, North America, Mediterranean area
Latrodectus mactans m.	South-, Central-, North America, Hawaii
L. m. tredecimguttatus	Mediterranean area, South Russia, Ethiopia, Arabia
L. m. cinctus	South- and East Africa, Ethiopia
L. m. menavodi	Madagascar
L. m. hasselti	South- and South-East Asia, New Zealand
L. pallidus	Russia, Syria, Israel, Iran, Libya
L. curacaviensis	from South Canada to Patagonia
Lycosa	South- and Central America, Mediterranean area
Phoneutria	Brazil (R. G. do Sul)
Trechona	South America
Xenesthis	Columbia, Venezuela, Panama
Megaphobema	Columbia
Pamphobeteus	South America
Lasiodora	Brazil
Teraphosa	Venezuela, Guyana
Phormictopus	Central America
Acanthoscurria	South America
Lithyphantes	Chile, Bolivia, Argentina, Brazil
Chiracanthium	Hawaii, Peru, Mediterranean area, Germany
Dendryphantes	Bolivia, Chile, Brazil

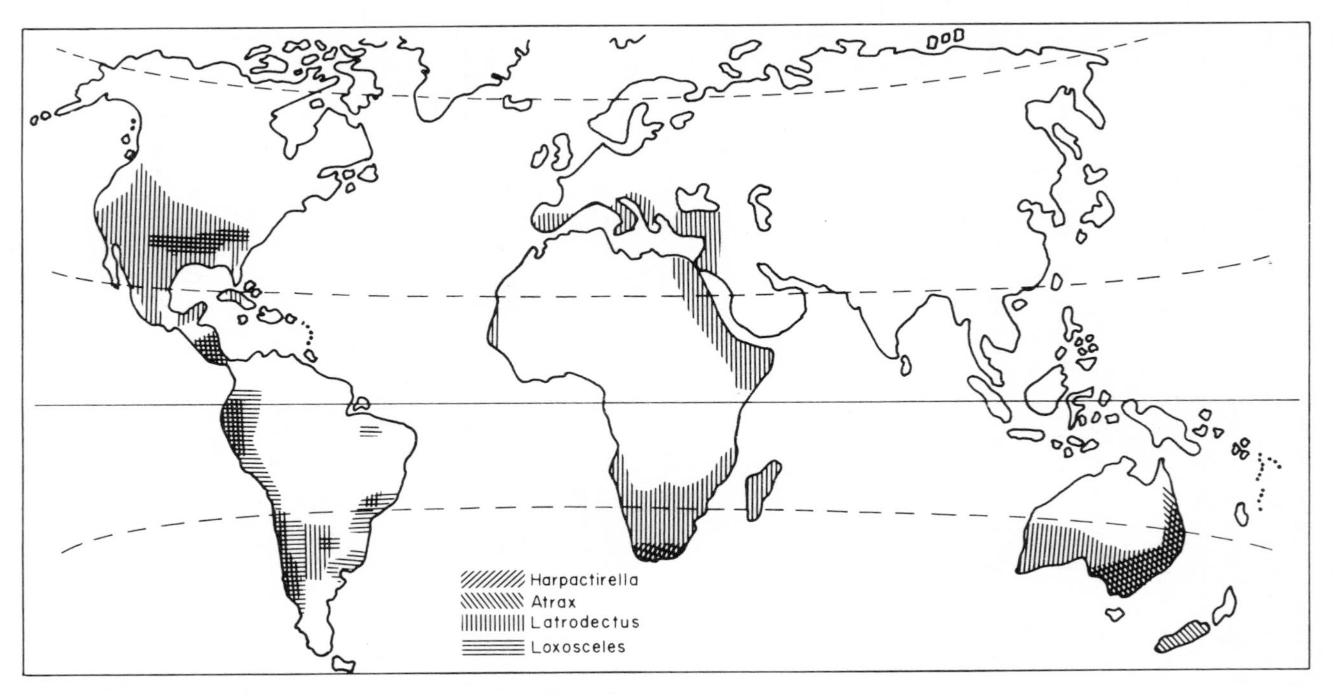

Fig. 6. Distribution of the most important venomous spiders

Envenomation

Poisoning always takes place as the result of a bite. Spiders usually are not aggressive unless they are frightened. With many species, however, other factors have to be taken into account, such as weather and season. It is well known that females are extremely dangerous from the mating season to the hatching of the youngs. Some data on toxicity and quantity of venom per animal are given in Table 8.

Table 8. Quantity of venom and lethal dose of some spider spp. (by Bücherl)

Species	Average quant. of venom	Max. quant. of venom	Lethal dose LD_{100} for a 20 g mouse	
	(mg)	(mg)	i. v.	s. c.
Orthognata				
Trechona venosa	1.00	1.70	0.030	0.070
Avicularia avicularia	1.30	6.50	–	–
Grammostola actaeon	3.70	5.20	0.490	1.150
Grammostola pulchripes	2.90	4.50	0.480	1.200
Acanthoscurria atrox	2.40	8.90	0.300	0.850
Acanthoscurria musculosa	2.30	4.20	0.210	0.450
Acanthoscurria sternalis	1.00	3.10	0.300	0.620
Acanthoscurria violacea	0.60	1.50	0.280	0.610
Eupalaestrus tenuitarsus	1.30	1.80	0.950	2.100
Eurypelma rubropilosum	2.00	6.00	0.350	0.850
Lasiodora klugi	2.40	3.60	0.640	1.200
Pamphobeteus roseus	1.60	3.00	0.850	1.700
Pamphobeteus sorocabae	0.80	2.80	0.700	1.500
Pamphobeteus tetracanthus	2.20	2.70	0.600	1.400
Labidognatha				
Loxosceles similis	0.70	1.5	0.130	0.250
Loxosceles rufipes	0.70	1.5	0.200	0.300
Latrodectus curacaviensis	0.60	1.3	0.170	0.240
Latrodectus geometricus	0.30	0.50	0.230	0.450
Lycosa erythrognatha	1.00	2.05	0.080	1.250
Phoneutria nigriventer	1.25	8.00	0.006	0.0134

Lycosa, Phoneutria, Loxosceles

According to all investigations reported so far, the most active toxin is that of *Phoneutria fera* (synonym: *Phoneutria nigriventer, armadeira,* banana spider). The bite is extremely painful, and death may occur within two to six hours. Once this time has been exceeded the chance of survival is very good.

Phoneutria toxin is a neurotoxin that acts on the central as well as the peripheral nervous system. The first symptoms appear within 10 to 20 minutes after the bite. In addition to the local burning pains at the site of the bite, which usually irradiate to the entire limb and even to the trunk, tachycardia, increase in blood pressure, vertigo, fever and sweating (mainly at the neck), loss of sight, nausea, vomiting, respiratory distress and paralysis are observed. Death is due to respiratory paralysis. Most fatal cases are among physically weakened persons and children; adults usually recover within one to two days. Accidents occur during the colder seasons, when spiders invade houses and workers handling clusters of bananas are frequently victims. In former times *Phoneutria fera* were inadvertently transported to Europe in banana boats, and accidents happened during unloading. In recent years these cases have become rare.

Treatment

Local pains are the main symptom, and certain features of other symptoms may be due to these pains. Hence pain has to be treated first. Novocain infiltration is useful. Peripheral vasodilation relieves pain, too, (e. g. by injection of nicotinamide). Morphine or other opium derivatives should be avoided, as they may potentiate the action of the venom. Antihistamines should be given before administering antivenin, since they minimize serum reactions. In severe cases 1 to 5 ampoules of antivenin must be given intramuscularly, followed by the same quantity intravenously.

A polyvalent antivenin is produced by the Instituto Butantan, Sao Paulo. The *Sero Antiaracidico Polivalente* is active against *Phoneutria, Lycosa* and *Loxosceles.*

The *Lycosa* toxins are cytotoxic. The bite is not accampanied by strong pains, but a local necrosis occurs. General symptoms are not observed. Less dangerous than famous is the venomous *Lycosa tarentula (Tarentula apuliae),* which is distributed in Italy, Sardinia and Spain, and the bite of which causes necrosis. In the 18th century a long and passionate dance was recommended for treatment: The Tarantella. Bites from this spider are very rare nowadays.

Loxosceles toxins possess cytotoxic and hemotoxic actions. Humans are obviously extremely sensitive to this venom. Spiders of the genus *Loxosceles* are widely distributed in temperate as well as tropical regions. 17 species are found in Africa, 2 in the warmer parts of Europe, and about 50 species have been described in North and South America. *L. rufescens* is well known in the Mediterranean area. It has been found in all the countries surrounding the Mediterranean Sea (including North Africa), Madagascar, Southern Russia, Burma, China, Japan and many islands in the Atlantic Ocean. It also has been found in the eastern USA, Mexico, Brazil and Paraguay. *L. reclusa* is native to the United States. *L. laeta* originated in the western parts of South America; they have been found in Argentina, Uruguay, Brazil, Chile, Peru and Ecuador.

From there it has spread into Central America as well as into the United States and the southern parts of Canada.

In recent years two *Loxosceles* species have gained importance in the South- and Southwest of the USA; they are responsible for 250 bites each year: *Loxosceles reclusa* and *Loxosceles laeta,* the

Table 9. Fatalities from *Loxosceles* bites (Table according to H. Schenone and G. Suarez in "Arthropod Venoms" (Ed. S. Bettini) p. 267)

Country	Years	No. of cases	Fatalities	Percentage
Chile	1873–1962	154	15	9.7
	1955–1966	200	7	3.5
	1966–1975	333	41	12.3
Argentina	1944–1966	17	3	17.6
Uruguay	1938–1953	29	2	6.9
Peru	1943–1952	31	5	16.1
	1950–1962	90	11	12.2
	1962–1969	52	5	9.6
Brazil	1954–1960	28	0	0.0
USA	1869–1968	126	72	4.8
	total	1060	161	15.2

latter originally native to South America. *L. laeta* is much more toxic than *L. reclusa.* For both of them the trivial names "Brown Recluse Spider" or "Violin Spider" (because of their body shape) are used. While *L. laeta* has emigrated via Central America into California, *L. reclusa* is native to Missouri, Kansas, Arkansas, Oklahoma and Texas. In recent years they have spread into Louisiana, Mississippi, Alabama, Georgia, Tennessee and Kentucky; some specimens of *L. reclusa* have been found in Arizona and California, too. The animals are found not only outdoors in protected places, but also in homes, frequently in closets where they eagerly consume insects, or in other places where they are undisturbed. The best protection against them is cleanliness. The bite of a female is much more dangerous than that of a male. Local pains are hardly initially present, but the site of the bite swells for some hours, and then a stinging-burning pain is observed. In serious cases this area, 3 mm to 3.5 cm in diameter, becomes dark red, and blisters are observed. The tissue finally turns black, and necrosis becomes evident. Healing makes slow progress and can take from 3 weeks (90%) to 5 months so that skin transplants may become necessary (10%). Corticosteroids have been successfully used. To avoid acute hemolytic renal failure, 10–15 g sodium bicarbonate per day should be given to increase the pH of the urine and to make the hemoglobin more soluble. Antiserum has been used with variable success.

In about 13% of all cases systemic symptoms are observed in addition to the local manifestations; hemolytic anemia, hemoglobinurea, hematuria and sensorial involvement appear within the first 24 hours after the bite. Body temperature may rise to 41°C, and in serious cases the victim may become comatose. The heart muscle, liver, lungs and kidneys are affected. These symptoms disappear within two weeks.

Chemically, the crude toxin consists of peptides with cytotoxic, neurotoxic and hemotoxic activities; the structure of these compounds is not yet clear.

Latrodectus

Latrodectus spp. (black widow spider) are not usually aggressive; they bite mostly in self defense but are able to do so several times in succession. Also, in these species the venom apparatus of the male is smaller than that of the female, and bites in humans occur only from the female.

The taxonomy of *Latrodectus* is unclear and contradictory. Here the trinomial systematics of *Levi* is used. *L mactans mactans* occurs in America, *L. m. tredecimguttatus* in Europe and North Africa, *L. m. menavodi* in Madagascar, *L.m. hasselti* in India and Australia. In addition to the *mactans* group there exist additional species: *L. geometricus* and *L. curacaviensis* in America, *L. hystrix* from the Yemen, *L. pallidus* from Palestine, Turkey and Southern Russia and *L. dahli* from Iran. Recently they have spread into many countries, and even in Belgium several specimens of *L. m. mactans* and *L. geometricus* have been found. They were probably carried in from elsewhere, but since they were found under natural conditions, it is possible that these species have already colonized Northern Europe.

The toxic effects of *Latrodectus* venom has been thoroughly investigated. The LD_{50} values differ greatly according to the species involved. Some are listed in the following table.

The surprisingly high tolerance of frogs to the venom is a result of its cutaneous respiration, which can overcome any defiency in pulmonary respiration produced by the venom.

Latrodectism has been observed in the warm zones of all continents, in Europe from the Mediterranean area to Northern France and Poland, in the Americas from Canada to Patagonia, and in practically all parts of Africa, Asia and Australia. Epidemics are particularly dangerous, when the number of these spiders increases

Table 10. LD_{50} values of *L. m. tredecimguttatus* venoms (according to Bettini and Maroli)

Animal	LD_{50} (mg/kg)
Frog	145 ± 32
Blackbird	5.9 ± 1.7
Pigeon	0.36 ± 0.7
Cockroach	2.7
Housefly	0.6
Guinea pig	0.075
Mouse	0.9
Rat	0.21

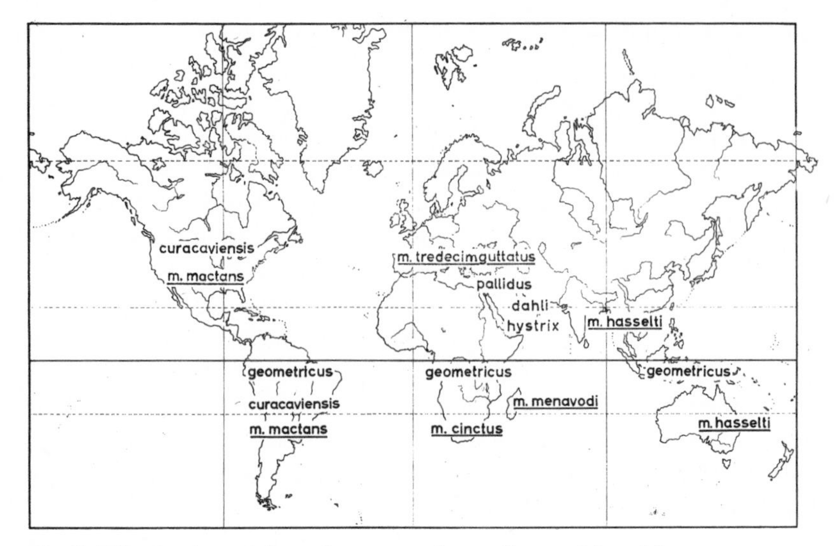

Fig. 7. Distribution of *Latrodectus spp.* (according to Maretic)

dramatically and hundreds or even thousand of animals are found in a rather limited area. Such events have been reported from the whole Mediterranean area and also from Brazil. The earliest report came from the year 866, when the troups of Emperor Ludwig were afflicted with this spider in Calabria/Italy; the last came from the Yugoslav province of Herzegovina from 1971–76. *L. m. tredecimguttatus* has been known to be dangerous since antiquity, however. The symptoms of latrodectism as well as fatalities have been known since the 4th century B. C., (Xenophon, Aristotle, Nikander of Kolophon, C. Plinius sec.).

Accidents usually occur in the fields when farmers are bitten during their work. In the last two decades, however, Latrodectus has also invaded towns and even cities. A preferred location are outdoor toilets where they can be found on the underside of the seats.

The poisoning is usually severe, and many fatalities are known. In 48 states of the USA in 14 years (1959–1973) 1,726 cases were reported, among them 55 fatal ones (3.2%). Similar statistics exist for other countries.

Symptomatology, Pathology and Treatment

Latrodectism in humans is a syndrome with a number of characteristic symptoms. The black widow produces a neurotoxin that acts on the spinal cord but also provokes many unspecific reactions, such as an intense feeling of anxiety of mortal fear.

The bite itself is usually very slight; about 60% of the patients observe no pains in connection with it. The first symptoms may arise with a latency of 10 minutes to 1 hour. A localized, slightly reddened goose-flesh area several millimeters in diameter is observed, which increases in size with time. Red "beams" corresponding to lymphatic pathways extend from the site of the bite to the lymphatic nodes, which are enlarged and ache. A few hours later a pallid area a few centimeters in diameter can be seen, surrounded by a red and blue circular border. The marks of the fangs are sometimes observed in the center, where anesthesia or hyperpathia occurs. Local necrosis usually occurs as the consequence of treatment with chemicals or incision. A scarlet-like rush accompanied by itching is very characteristic on the fourth day after the bite.

The general condition is usually poor. Pain spreads from the lymphatic nodes to the abdomen, waist, thighs and the lower extremities and may persist for 20 hours or more. Feeling of pressure in the chest may occur as well as tremor of the whole body. Sweating and lachrimation are connected with muscular contractions of the face. A painful grimace is typical: The *facies latrodectismica.*

The principal cardiac symptoms are a moderate tachycardia (up to 120/min) observed during the first hours, followed by a marked bradycardia. The main symptoms come from the neurotoxic effects of the venom.

The cerebral and the vegetative systems are affected. Acute psychoses may develop; dread and fear of death are frequent symptoms together with restlessness and crying. Such psychic disorders may last for three days. Also headache is frequently observed. Increase of all secretions is typical: perspiration, hypersalivation, rhinitis, lachrimation, increase of the secretions from the gastrointestinal tract are characteristics of latrodectism.

The acute symptoms of latrodectism usually last two to three days in untreated patients, the total length being less than one week.

Hospitalization therefore usually is required in patients younger than 16 and over 60 years, patients with hypertensive heart disease and in those with signs of severe intoxication. Patients properly treated with antivenom will recover rapidly.

On the other hand, convalescence lasts a long time. Weakness, fatigue, pains, headaches, sleeplessness and impotence may persist for several months. Complications are not frequent, although in a number of patients with preexisting conditions such complications have been observed: Hypertension may provoke cerebral insult, heart failure, kidney failure or severe emphysema.

The prognosis is generally good. The statistics differ very much concerning fatalities. From 618 cases recorded in the USA, 38 (6.2%) were fatal. In another 177 cases from Yugoslavia, no deaths were observed. Therapy should avoid aspecific methods. Local treatment such as cauterization, incision or application of chemicals are completely useless and may cause secondary infections and other problems more dangerous than the venom itself. Alcohol is contraindicated. Neither morphine nor barbiturates should be applied for alleviation of pain, since they must be used in heavy doses that could have negative effects. Infiltration with 1% procaine may be performed.

A number of different preparations have been used by several authors, but the results are not consistent. Without any doubt the best treatment utilizes antivenins, which is prepared in the USA by Merck, Sharp and Dohme Co.; in South Africa by the South African Institute for Medical Research in Johannesburg, and in Italy by the Instituto Superiore di Sanita in Rome. 2–5 ml of antivenin usually suffices. The action of antivenin can be potentiated by prior injection of calcium intravenously; dramatic recovery within a short time can be observed.

In other cases, for example if antivenin cannot be applied, 10 ml of methocarbamol (Robaxin) i. v. over a five minute period followed by another 5 to 10 ml in a drip of 150 ml of 5% dextrose in water may be infused. If relief ensues, 500 mg Robaxin every 6 hours should be given for 24 hours. In the very rare instances of local pains, infiltration with 1% procaine can be done locally.

Atrax, Chiracanthium

The most dangerous spiders of Australia are the nine species of the genus *Atrax,* among which *Atrax robustus* is the most dangerous. The main distribution area is New South Wales and the adjacent Blue Mountains. Even in the city of Sydney deaths are occasionally caused by this spider. The female is 4 cm in length (8 cm including legs); the male is smaller. The strong fangs may inflict a wound similar to a snake bite. Practically all fatal cases so far unambigously attributed (within a radius of 180 km of Sydney) have been caused by males. Fortunataly, fatalities are rare, 11 cases in all since 1927.

The *Atrax robustus* bite is extremely painful because of the large fangs and the low pH of the venom. Pain may persist for several days, but no necrosis is observed. Systemic symptoms usually develop within minutes after the bite. The classical syndrome consists of nausea and vomiting, abdominal pain and diarrhea, sweating, salivation, lachrimation and dyspnea. Blood pressure is elevated; spasms may occur but no paralysis. A profound coma may last for many hours. Confusion and hypoxia may result in an asphyxial death. Death may occur within one hour after the bite or even after 30 hours. Occasional improvements after a few hours should not lead to the assumption that the worst is over.

Reports of bites from species other than *A. robustus,* e. g. *A. formidabilis,* are rare, but as far as we know poisoning exhibits the same symptoms. No antivenin is available for treatment. Prompt application of an arterial tourniquet (*not* a venous tourniquet) is necessary, and the patient should be immediately transported to an intensive care unit. The tourniquet should *not* be released in transit even if there is a risk of ischemic tissue alteration. The envenomation must be treated symptomatically, and the patient should stay in the hospital at least 24 hours after the last symptoms have disappeared.

Harpactirella lightfooti, the South African "Tarantula" or "baboon spider" occasionally causes envenomations with local reactions (redness, pain) and sometimes collapse. No severe cases or fatalities have been reported.

Chiracanthium spp. are widely distributed, and poisonings have been reported since the 18th century. *Chiracanthium punctorium* is indigenous to Europe; it has been reported in Italy, Yugoslavia,

Germany, Switzerland and France. The body of the male is 12 mm. of the female 15 mm in length. The diameter including legs is 3 cm. The females possess extremely strong fangs. The animals may be found in moist meadows during summertime, in sheds, storage houses or barns in winter. They build their cocoons in June and July, which may have the size of a dove or chicken egg. Mating time is in June and the eggs are laid in August. The cocoon with the eggs is guarded by the female, which is then very aggressive. The spiderlings hatch in Fall; they hibernate in funnel-like webs in protected places.

Symptoms of the bite include localized stinging and extreme burning pain; the site of the bite is discolored reddish-blue for some days and is swollen. The swelling and pain may be limited to the very site of the bite, but it may also extend into the whole limb. The lymph nodes are enlarged. Sometimes the local symtoms resemble a small viper bite. After a while the pain changes to itching and paresthesia. Besides the local reactions, systemic symptoms occur. Shivers and an increase of body temperature to 38.5°C, nausea, vomiting and headache have been observed; oppression in the chest and collapse have been reported, too.

The sytemic symptoms disappear within a few days; they can be treated symptomatically in case of collapse. In some cases lesions and necrosis at the site of bite were reported, which had to be treated surgically. Local treatment by incision, cauterization, chemicals or cryotherapy are not indicated; damage caused from such "prehistorical" treatment is by far more severe than those of the bite. No serious cases or fatalities have been observed.

Other species that cause essentially the same symptoms are *C. inclusum* and *C. mildei* (USA), *C. mordax* and *C. longimanus* (Australia) and *C. japonicum* (Japan).

Chemistry

Chemical studies on spider venoms are in progress, but the results have not yet provided great detail. We know that the crude venoms are complex mixtures. They contain enzymes such as phosphatase, collagenase, esterase, hyaluronidase, phospholipases A, C and D, and protease. Furthermore specific toxins with cardiotoxic or neurotoxic activity can be found; they are peptides of different chain length, but mostly with a molecular weight above 6,000. The primary structures of these toxins have not yet been established. And finally, biogenic amines such as histamine or serotonin have been detected.

References

Bücherl, W.: In "Venomous Animals and their Venoms" (Ed.: W. Bücherl and E. Buckley),, Vol. III, chapter 51, Academic Press, New York, 1971

Schenberg, S., Pereira Lima, F. A.: ibid, chapter 52

Maretic, Z.: ibid, chapter 53

Maretic, Z.: Med. Klinik *57,* 1576 (1962)

Habermehl, G.: Naturwissenschaften *61,* 368 (1974)

Maretic, Z.: Bull. Brit. Arach. Soc. *3,* 126 (1975)

Bettini, S. (Ed.): Arthropod Venoms, Springer Verlag, Berlin, Heidelberg, New York, 1978

Maretic, Z.: Natur und Museum *95,* 124 (1965).

3.2 Myriapoda

3.2.1 Chilopoda (Centipedes)

The class *Chilopoda* is broadly distributed and contains about 3,000 species. They have a long, slender body about 1 cm in diameter and up to 30 cm in length. They are distributed in four orders: *Sentigeromorpha, Lithobiomorpha* (both with 15 pairs of legs), *Scolopendromorpha* (21–23 pairs of legs) and *Geophilomorpha* (33–177 pairs of legs). Centipedes live under stones, in dark places or caves, under bark and leaves. For predation they possess a pair of venom glands with forcipules at the first trunk segment.

Bites of the larger species may be of medical significance, although the many reports of poisonings are the subject of some doubt. Irritation, swelling of the bite site and ulceration have occasionally been observed, but could have been due to secondary infections. Fatalities were reported in early papers, but in recent publications they could not be confirmed. In any case, for small mammals the venom is highly active; the mean lethal doses can be seen in the following Table 11. The venom acts as a neurotoxin and produces corresponding symptoms: tachypnea, profuse se wating, dizziness, nausea and vomiting, convulsions and death by respiratory paralysis.

There are no recent publications on the chemistry and biochemistry of these venoms. Immunological studies are also not yet available.

Table 11. LD_{50} of venom of scolopendromorphs (mg/mouse, 20 g). (from A. Minelli in "Arthropod Venoms" (Ed. S. Bettini), p. 81)

Species	Body length of centipede (cm)	*Median lethal dose*	
		intra-venously	intra-muscularly
Scolopendra viridicornis	16–19	0.030	0.250
S. subspinipes	11–18	0.047	1.2
Otostigmus scabricauda	6–7	0.012	0.070
Cryptops iheringi	6–9	0.150	0.340
Scolopocryptos ferrugineus	5–7	0.160	0.390

In addition to these active venoms, the *Chilopoda* possess a rich variety of defense secretions as weapons against ants, beetles and other enemies. They are produced and secreted in the ventral and coxal glands. Some of them are sticky, some luminescent and others possess an intense odor. Some of these secretions contain compounds generating hydrogen cyanide, such as mandelonitrile and benzoylcyanide.

3.2.2 Diplopoda (Millepedes)

The second class of the *myriapoda* are the millepedes. They are a very ancient group of animals dating back to Devonian times, comprising about 7,500 species. They have developed a very effective defense mechanism, consisting of secretions that are excreted from glands located in the segments when the animal is endangered. Chemically they consist of relatively simple compounds; a review is given in the following Table 12.

Table 12. Chemical components of defensive secretions of millepedes

	Chemical components	Millepedes in which components are found[a]
I	Hydrogen cyanide	37–41, 43–55
II	Formic acid	52
III	Acetic acid	52
IV	Isovaleric acid	52, 54
V	Myristic acid	54
VI	Stearic acid	54
VII	Hexadecyl acetate	30
VIII	Δ^9-Hexadecenyl acetate	30
IX	Δ^9-Octadecenyl acetate	30
X	*trans*-2-Dodecenal	8
XI	Benzaldehyde	37, 39, 40, 42–44, 48–50, 52, 54
XII	Benzoic acid	44, 48, 52, 54
XIII	Mandelonitrile	37, 39, 44, 54
XIV	Mandelonitrile benzoate	44, 52, 54
XV	Benzoyl cyanide	37, 39, 54
XVI	Phenol	48, 49
XVII	*o*-Cresol	34
XVIII	*p*-Cresol	36
XIX	Guaiacol (= 2-methoxyphenol)	42, 43, 48
XX	1,4-Benzoquinone	28, 35
XXI	2-Methyl-1,4-benzoquinone	2–17, 19–21, 24–27, 29–33
XXII	2-Methyl-3-methoxy-1,4-benzoquinone	2–6, 10, 11, 13–20, 22, 23, 25, 26, 30–34
XXIII	2,3-Dimethoxy-1,4-benzoquinone	3, 35
XXIV	5-Methyl-2,3-dimethoxy-1,4-benzoquinone	3
XXV	2-Methyl-1,4-hydroquinone	33
XXVI	2-Methyl-3-methoxy-1,4-hydroquinone	33
XXVII	*Glomerin* (= 1,2-dimethyl-4(3H)-quinazolinone)	1
XXVIII	*Homoglomerin* (= 1-methyl-2-ethyl-4(3H)-quinazolinone)	1
XXIX	*Polyzonimine*	56
XXX	*Nitropolyzonamine*	56

[a] Numbers refer to millepedes listed in Table 13.

Table 13. Millepedes in which chemical component of defensive secretion have been identified

Milliped	Chemical components present in the defensive secretions
Subclass Pentazonia	
Order Glomerida	
1. *Glomeris marginata*	XXVII, XXVIII
Subclass Helminthomorpha	
Order Spirobolida	
2. *Chicobolus spinigerus*	XXI, XXII
3. *Epibolus pulchripes* (= *Metiche tanganyicense*)	XXI, XXII, XXIII, XXIV
4. *Floridobolus penneri*	XXI, XXII
5. *Narceus annularis*	XXI, XXII
6. *N. gordanus*	XXI, XXII
7. *Pachybolus laminatus*	XXI
8. *Rhinocricus insulatus*	X, XXI
9. *R. varians*	XXI
10. *Rhinocricus* sp.	XXI, XXII
11. *Trigoniulus lumbricinus*	XXI, XXII
Order Spirostreptida	
12. *Aulonopygus aculeatus*	XXI
13. *Archispirostreptus gigas*	XXI, XXII
14. *A. tumuliporus*	XXI, XXII
15. *Cambala hubrichti*	XXI, XXII
16. *Collostreptus fulvus*	XXI, XXII
17. *Doratogonus annulipes*	XXI, XXII
18. *Orthoporus conifer*	XXII
19. *O. flavior*	XXI, XXII
20. *O. ornatus (= O. punctilliger)*	XXI, XXII
21. *Peridontopyge aberrans*	XXI
22.*P. coani*	XXII
23. *P. rubescens*	XXII
24. *P. tachoni*	XXI
25. *Prionopetalum frundsbergi*	XXI, XXII
26. *P. tricuspis*	XXI, XXII
27. *Rhapidostreptus virgator* (= *Sprirostreptus v.*)	XXI
28. *Spirostreptus castaneus*	XX
29. *S. multisulcatus*	XXI

Millepedes	Chemical components present in the defensive secretions
Order Julida	
30. *Blaniulus guttulatus*	VII, VIII, IX, XXI, XXII
31. *Chromatoiulus unilineatus* (= *Brachyiulus u.*)	XXI, XXII
32. *Cylindroiulus londinensis* (= *C. teutonicus*)	XXI, XXII
33. *Ommatoiulus sabulosus* (= *Archiulus s.*)	XXI, XXII, XXV, XXVI
34. Oriulus delus	XVII, XXII
35. *Uroblaniulus canadensis*	XX, XXIII
Order Callipodida	
36. *Abacion magnum*	XVIII
Order Polydesmida	
37. *Apheloria corrugata (= A. coriacea)*	I, XI, XIII, XV
38. *A. kleinpetri*	I
39. *A. trimaculata*	I, XI, XIII, XV
40. *Astrodesmus laxus*	I, XI
41. *Cherokia geordiana*	I
42. *Euryurus australis*	XI, XIX
43. *E. leachii*	I, XI, XIX
44. *Gomphodesmus pavani*	I, XI, XII, XIII, XIV
45. *Harpaphe haydeniana* (= *Leptodesmus h.*)	I
46. *Motyxia sequoiae* (= *Luminodesmus s.*)	I
47. *Nannaria* sp.	I
48. *Orthomorpha coarctata*	I, XI, XII, XVI, XIX
49. *Oxidus gracilis* (= *Fontaria g.*)	I, XI, XVI
50. *Pachydesmus crassicutis*	I, XI
51. *Pleuroloma flavipes*	I
52. *Polydesmus collaris*	I, II, III, IV, XI, XII, XIV
53. *Pseudopolydesmus branneri*	I
54. *P. serratus*	I, IV, V, VI, XI, XII, XIII, XIV, XV
55. [= *Polydesmus (Fontaria) virginienses*]	I
Order Polyzoniida	
56. *Polyzonium rosalbum*	XXIX , XXX

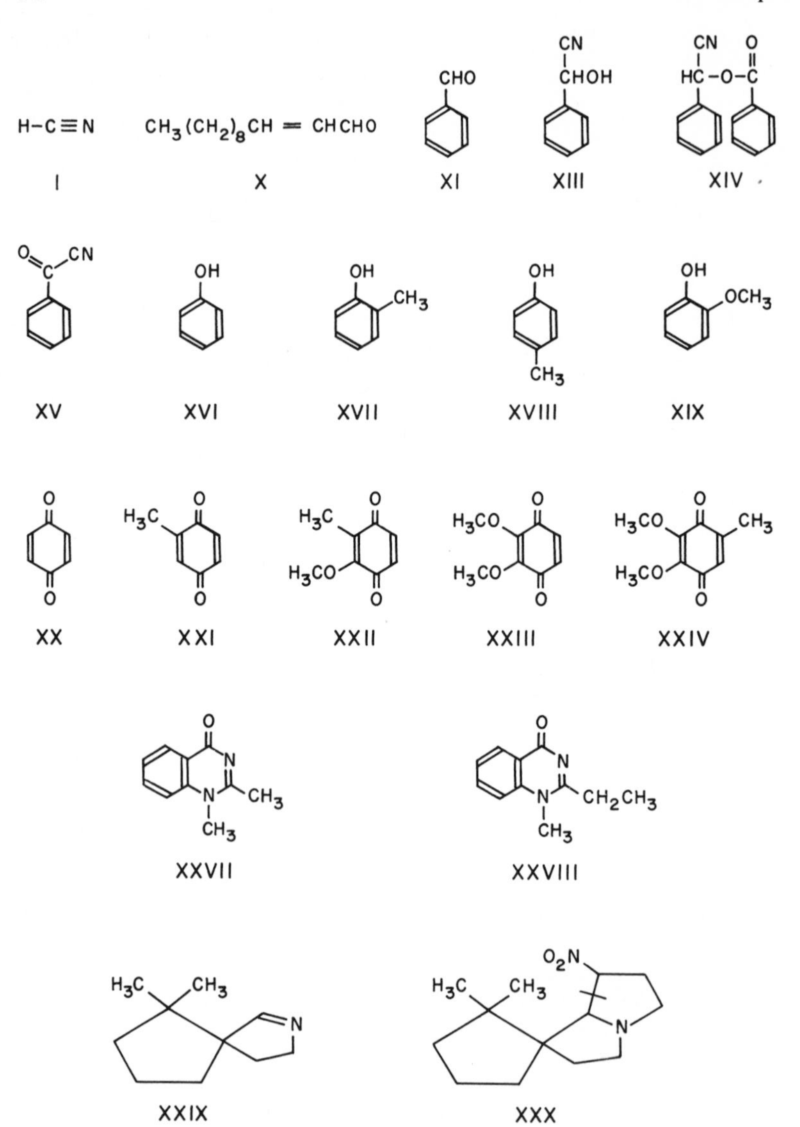

Fig. 8. Chemical components of defensive secretions of millepedes. Roman numerals correspond to compounds named in Tables 12 and 13.

These low molecular weight benzene derivatives and nitrogen-containing alkaloids are frequently mixed with proteinaceous material that hardens quickly on exposure to air and sticks to attacking animals. The individual compounds have different modes of action: Hydrogen cyanide produced, for example, from mandelonitrile is toxic; others may act as repellents.

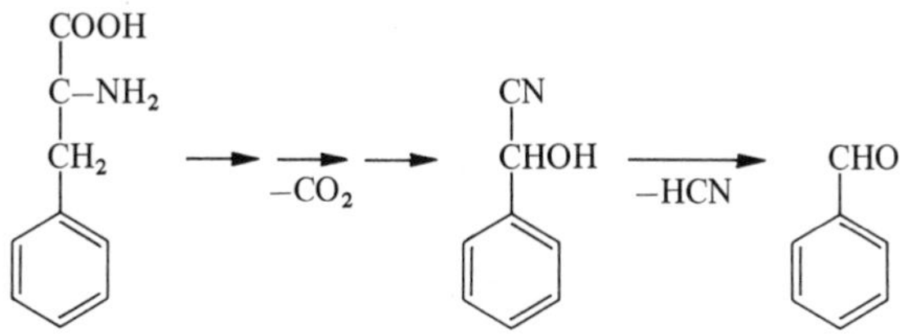

COOH
C–NH_2
CH_2 → $\xrightarrow{-CO_2}$ → CN
CHOH $\xrightarrow{-HCN}$ CHO

Of particular interest are the alkaloids *glomerin* and *homoglomerin* from *Glomeris marginata.* They have a paralytic action on mice as well as on spiders. *Homoglomerin* is biosynthesized from anthranilic acid:

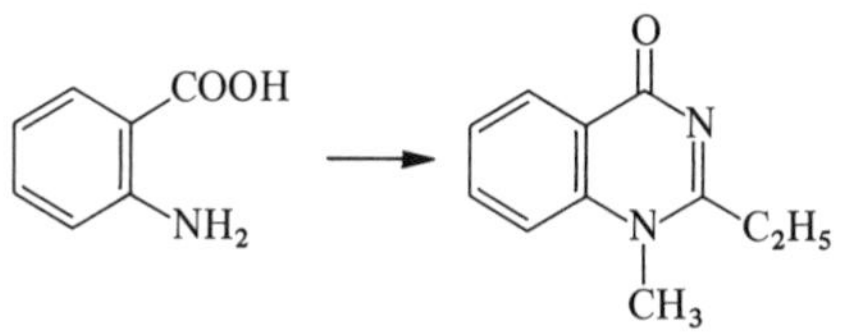

COOH
NH_2 → O
N
N C_2H_5
CH_3

Quite unique also are polyzonimine and nitro-polyzonimine from Polyzonium spp.

Many of these low molecular weight compounds, especially phenoles and quinones, are powerful antibiotics, and it may well be that this property is also one of the biologic significances of these substances.

3.3 Hexapoda (Insects)

3.3.1 Dermaptera (Earwigs)

3.3.1.1 Forficulidae (Earwigs)

The earwigs are one of the small orders of insects and comprise about 900 species. They are widely distributed and frequent in most warm regions of the world and in Europe. Curiously they are rare in the United States. Earwigs are brown to black, flat, slender and between 0.5 and 4 cm in length. Some species possess wings, others do not. Most of them feed on plant material, some, however, on eggs or larvae of other insects. They are not actively venomous, but as in the case of *Diplopoda* they possess defense substances that are excreted from an abdominal gland. The common earwig, *Forficula auriculosa,* indigenous to Central Europe, contains in these odor glands Toluquinone and ethylbenzoquinone. In emergency situations theses animals can deliver 20 mg all at once.

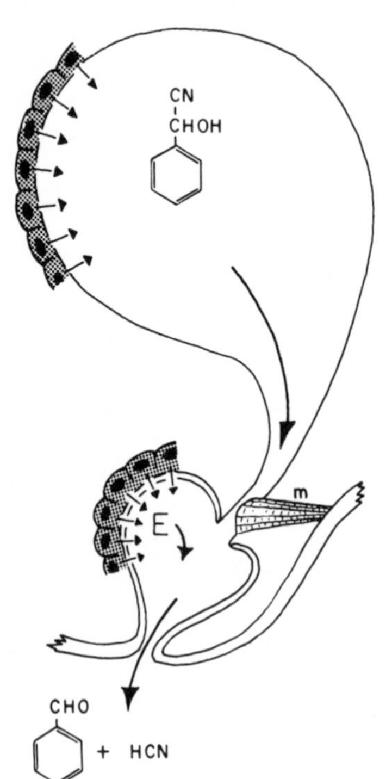

Fig. 9. Diagram of the two-chambered cyanogenetic glandular apparatus of the polydesmoid milliped, *Apheloria corrugata.* The inner compartment (reservoir) stores mandelonitrile, while the smaller compartment (vestibule) contains an enzyme *E* that catalyzes the breakdown of mandelonitrile into hydrogen cyanide and benzaldehyde. A muscle *m* operates the valve between the two compartments

3.3.2 Hemiptera: Heteropteroidea (Bugs)

The Hemiptera comprise about 55,000 known species throughout the world. They are passively venomous animals; the secretions they produce are only for defense, and they do not pose a danger to humans unless they are ingested or carelessly handled. Most of these defense secretions are malodorous; some of them irritate the skin or the mucous membranes. No poisoning has yet been observed.

Table 14. Distribution, within hemipteran families, of various classes of organic compounds found in their defensive secretions

		Compounds												
Ryneota	No. of species examined	Alkanes	Alkanals	Alkenals	Alkanones	Alkenones	Alkanols	Alkanol esters	Alkenol esters	Dicarbonyl compounds	Carboxylic acids	Steroids	Aromatic compounds	Miscellancous compounds
Heteroptera														
Belastomatidae	2								×					
Cimicidae	1		×	×	×									
Coreidae	23		×	×			×	×		×	×			
Corixidae	2									×				
Cydnidae	2	×	×	×					×	×				×
Gelastocoridae	1									×				
Lygaeidae	3			×				×		×		×		
Miridae	1		×	×									×	
Naucoridae	1												×	
Notonectidae	1													
Pentatomidae	25	×	×	×	×	×			×	×				
Plataspidae	1	×												
Pyrrhocoridae	1	×	×	×						×				×
Reduviidae	3										×			
Homoptera														
Aphidae	1											×		

The Hemiptera are divided into two sub-orders, the Heteroptera and the Homoptera. The chemistry of the defense secretions in given in the following Tables 14, 15, 16.

Table 15. Compounds used for defense by aquatic hemipteran species

Family, genus and species	Compound
Belastomatidae	
Lethocerus cordofanus	*trans*-hex-2-enyl acetate
L. indicus	*trans*-hex-2-enyl acctate *trans*-hex-2-enyl butyrate
Corixidae	
Corixa dentipes	*trans*-4-oxo-hex-2-enal
Sigara falleni	*trans*-4-oxo-hex-2-enal
Gelastocoridae	
Gelastocoris oculatus	*trans*-4-oxo-hex-2-enal
Naucoridae	
Ilyocoris cimicoides	*p*-hydroxybenzaldehyde methyl-*p*-hydroxybenzoate
Notonectidae	
Notonecta glauca	*p*-hydroxybenzaldehyde methyl-*p*-hydroxybenzoate

Table 16. Compounds used for defense by some terrestrial hemipteran species

Family, genus, and species	Compound
Cimicidae	
Cimex lectularius	Acetaldehyde, *trans*-hex-2-enal, *trans*-oct-2-enal, butan-2-one
Coreidae	
Acanthocephala declivis, A. femorata, A. granulose	*trans*-hex-2-enal, acetic acid
Amorbus alternatus	hexanal, hexyl acetate, hexanol, acetic acid
A. rhombifer	butanal, hexanal, butyl butyrate, hexyl acetate, acetic acid
A. rubiginosus, Aulacosternum nigrorubrum	hexanal, hexyl acetate, hexanol, acetic acid
Leptoglossus clypeatus, L. oppositus	hexanal, acetic acid

Family, genus, and species	Compound
Cimicidae	
Cimex lectularius	Acetaldehyde, *trans*-hex-2-enal, *trans*-oct-2-enal, butan-2-one
Libyaspis angolensis	propanal, butanal, *trans*-dec-2-enal, *trans*-hex-2-enal, *trans*-4-oxo-hex-2-enal
Mictis caja	hexanal, butyl butyrate, hexyl acetate, hexanol, acetic acid
Mozena lunata, M. obtusa	hexanol, hexyl acetate, acetic acid
Pternistria bispina	
(adult)	butanal, hexanol, hexyl acetate. hexyl butyrate
(larvae)	*trans*-hex-2-enal, *trans*-oct-2-enal, *trans*-4-oxo-hex-2-enal
Cydnidae	
Macroscytus sp.	dodecane, tridecane, *trans*-4-oxo-hex-2-enal, *trans*-oct-2-enyl acetate, *trans*-dec-2-enyl acetate
Scaptocoris divergens	propanal, propenal, butenal, pentenal, hex-2-enal, heptenal, octenal, furan, methyl furan, toluquinone
Lygaeidae	
Caenocoris nerii	adigoside, nerigoside, neritaloside, odoroside A, strospeside
Oncopeltus fasciatus	
(adult)	hex-2-enal, oct-2-enal, hexa-2,4-dienal, oct-2,4-dienal, hex-2-enyl acetate, oct-2-enyl acetate, hexa-2,4-dienyl acetate, oct-2,4-dienyl acetate
Miridae	
Leptopterna dolabrata	acetaldehyde, *trans*-oct-2-enal
Pentatomidae	
Aelia fieberi	*trans*-oct-2-enal, *trans*-dec-2-enal
Apodiphus amygadi	dodecane, tridecane, hexanal, octanal, decanal, hexenal, 4-oxo-hex-2-enal, 4-oxo-oct-2-enal

Table 16 (continued)

Family, genus, and species	Compound
Cimicidae	
Cimex lectularius	Acetaldehyde, *trans*-hex-2-enal, *trans*-oct-2-enal, butan-2-one

Family, genus, and species	Compound
Biprorulus bibax	undecane, dodecane, tridecane, pentadecane, *trans*-hex-2-enal, *trans*-dec-2-enal, *trans*-hex-2-enyl acetate, *trans*-dec-2-enyl acetate
Dolycoris baccarum	*trans*-hex-2-enal, *trans*-oct-2-enal, *trans*-dec-2-enal
Eurydema rugosa, E. pulchra	tridecane, *trans*-hex-2-enal
Halys dentata	hexanal, octanal, dec-2-enal, butan-2-one
Musgraveia sulciventris	undecane, dodecane, tridecane, pentadecane, *trans*-hex-2-enal, *trans*-oct-2-enal, *trans*-dec-2-enal, *trans*-4-oxo-hex-2-enal, *trans*-4-oxo-oct-2-enal, *trans*-oct-2-enyl acetate tridecenyl butyrate, tetradecenyl butyrate
Nezara viridula	tridecane, *trans*-hex-2-enal, *trans*-dec-2-enal
Nezara viridula var *smaragdula*	undecane, dodecane, tridecane, *trans*-prop-2-enal, *trans*-but-2-enal, *trans*-hex-2-enal, *trans*oct-2-enal, *trans*-dec-2-enal, *cis*-dec-2-enal, butan-2-one, hexan-2-one, octan-2-one, hex-2-en-4-one, *trans*-4-oxo-hex-2-enal, *trans*-4-oxo-oct-2-enal, *trans*-hex-2-enyl acetate, *trans*-oct-2-enyl acetate, *trans*-dec-2-enyl acetate
Tessaratoma aethiops	tridecane, *trans*-hex-2-enal, *trans*-oct-2-enal, *trans*-4-oxo-hex-2-enal, *trans*-oct-2-enyl acetate
Vitellus insulanis	undecane, dodecane, tridecane, *trans*-dec-2-enal, *trans*-4-oxo-hex-2-enal
Plataspidae	
Ceratocoris cephalicus	tridecane

Family, genus, and species	Compound
Cimicidae	
Cimex lectularius	Acetaldehyde, *trans*-hex-2-enal, *trans*-oct-2-enal, butan-2-one
Pyrrhocoridae	
Dysdercus intermedius	
(adult)	acetaldehyde, octanal, *trans*-hex-2-enal, *trans*-oct-2-enal, terpene hydrocarbon
(larvae)	dodecane, tridecane, pentadecane, hexanal, *trans*-hex-2-enal, *trans*-oct-2-enal, *trans*-4-oxo-hex-2-enal, *trans*-4-oxo-oct-2-enal

3.3.3 Coleoptera (Beetles)

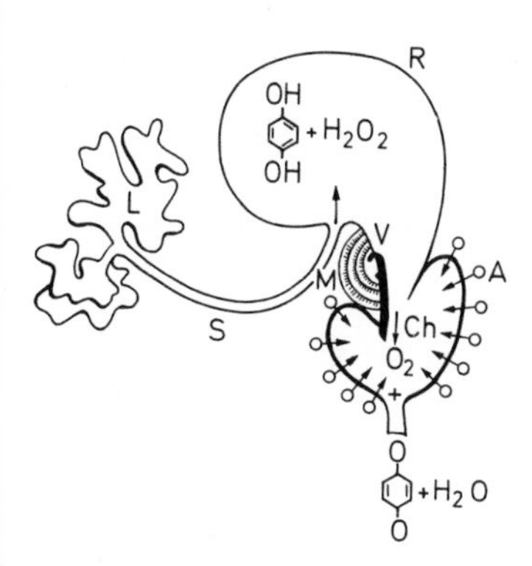

L lobes with glandular cells
S collective duct
R reservoir, Pygidial vesicle
V valve plate
M muscle
A annex glands
Ch chitin capsule

Fig. 10. Shooting mechanism of a "bombarding beetle". The glandular cells secrete hydroquinone, tolu-hydroquinone and hydroperoxide. This secretion mixture is stored in the Pygidial vesicle which at its end is separated from the real reaction chamber, the chitin capsule, by a muscular valve. When discharging, the beetle opens the valve to let the pygidial secretions enter into the chitin capsule where they undergo an explosion-like reaction. Quite probably we may assume that catalase is present here, for a stable solution of the vesicle content may react with it, and vice versa it is possible by means of the annex gland secretion to decompose a solution of 10% hydroquinone and tolu-hydroquinone in 25% hydroperoxide with strong development of gas. This gas delivers the pressure needed for emitting and spraying the quinones (according to H. Schildknecht)

The order *Coleoptera* comprises about 200 families with more than 250000 species. They use the secretions of the scent glands for defense only, and no poisoning of a human has been observed.

The chemistry of these venoms is more complex with respect to the variety of compounds. Some of these animals have developed very special mechanisms, for example the *Brachinidae* (assigned to the Carabidae). In these "bombarding beetles" *(Brachynus creptitans)* the defense substance, consisting of benzoquinone and toluquinone, is not simply secreted: The secretion consists of a 10% solution of the corresponding hydroquinones in 25% hydrogen peroxide. The reaction of both components is prevented by means of a stabilizer. When the animal must defend itself part of the solution is brought from the pygidial gland into an "explosion chamber", where the mixture is enzymatically destabilized. The hydrogen peroxide is decomposed, the hydroquinones are oxidized to the corresponding quinones and then projected a few inches with

Table 17. Distribution within *coleopteran* families of various classes of organic compounds present in their defense toxins

Coleoptera	No. of species examined	Compounds											
		Aliphatic acids	Aliphatic aldehydes	Aliphatic esters	Hydrocarbons	Aromatic acids	Aromatic aldehydes	Aromatic esters	Quinonoid compounds	Steroidal compounds	Terpenoid compounds	Alkaloids	Miscellaneous
Alleculidae	1								×				
Cantharidae	1	×											
Carabidae	153	×	×	×	×	×	×		×				×
Cerambycidae	4				×		×				×		×
Chrysomelidae	4						×	×					
Coccinellidae	14											×	
Dytiscidae	20					×	×	×		×			×
Gyrinidae	9		×								×		×
Meloidae	?										×		
Silphidae	3												×
Staphylinidae	14	×	×		×				×		×	×	×
Tenebrionidae	108	×	×		×				×				×

The location of toxins of coleopteran species may be seen in Table 18.

a loud crack. Since the beetles are able to "discharge" several times in succession, this is a very effective weapon (Fig. 10).

A survey of the compounds found in *Coleoptera* is given in the following Table 17.

Table 18. Location of toxin producing glands in *Coleoptera*

	Pygidial gland	Accessory pygidial gland	Thoracic gland	Tergal gland	♂ accessory gland	Mandibular gland
Alleculidae	×		(×)			
Cantharidae *Cantharis rustica* (adult and larva)				×		
Carabidae	×	(×)				
Cerambycidae						×
Chrysomelidae				×		
Dytiscidae	×	(×)	×			
Gyrinidae	×					
Haliplidae	×					
Lagriidae		×				
Meloidae					×	
Noteridae	×	–	–			
Staphylinidae	×					
Tenebrionidae	×		(×)			

3.3.3.1 Carabidae

The toxins of the *Carabidae* are listed in Table 19:

Table 19. Defensive toxins of the *Carabidae*

Carabidae	Components identified in the secretion
Abacomorphus asperulus	Angelic acid; formic acid; methacrylic acid; tiglic acid
Abax ater, A. oralis, A. parallelus	Methacrylic acid; tiglic acid

Table 19 (continued)

Carabidae	Components identified in the secretion
Acinopus sp., *Agonum assimilis*	Formic acid
A. (Idiochroma) dorsalis	Decane; formic acid; salicylic acid methyl ester, undecane
A. marginatum, A. mostum, A. sexpunctatum, A. viduum	Formic acid
Amara familaris, A. similata	Decane; methacrylic acid; tiglic acid; tridecane; undecane
Amblytelus curtus, Anisodactylus binotatus	Formic acid
Anthia thoracica	Acetic acid; formic acid; tiglic acid; isovaleraldehyde
Arthropterus sp.	*p*-ethylquinone; *p*-toluquinone
Asaphidion flavipes	Salicylaldehyde; valeric acid
Badister bipustulatus	Formic acid
Bembidion andreae, B. lampros	isobutyric acid; isovaleric acid
B. quadrigattatum	Salicylaldehyde, valeric acid
B. quadrimagulatum	*m*-cresol.
Brachynus crepitans, B. explodens, B. sclopeta	*p*-benzoquinone; *p*-toluquinone
Broscus cephalotes	isobutyric acid; isovaleric acid
Calathus fuscipes, C. melanocephalus	Formic acid
Callisthenes luxatus	Methacrylic acid; salicylaldehyde
Callistus lunatus	*p*-benzoquinone; *p*-ethylquinone; *p*-toluquinone
Calosoma affini, C. alternans sayi	Salicylaldehyde
C. externum	Methycrylic acid; salicylic acid
C. macrum	Salicylaldehyde
C. marginalis	Methacrylic acid; salicylaldehyde
C. parricollis	Salicylaldehyde
C. peregrinator	Methacrylic acid; salicylaldehyde
C. prominens	Salicylaldehyde
C. schayeri	Caproic acid; methacrylic acid; salicylaldehyde
C. scrutator	Methacrylic acid; salicylic acid
C. sycophanta	Methacrylic acid; salicylaldehyde; tiglic acid

Carabidae	Components identified in the secretion
Cychrus rostratus	Tiglic acid
Diachromus germanus, Diaphoromenus edwardsi, Dicaelus dilatatus, D. purpuratus, D. splendidus, Dichirotrichus obsoletus, Dicrochile brevicollis, D. goryi, Drypta dentata	Formic acid
Elaphrus ripareus	isobutyric acid; isovaleric acid
Eudalia macleayi	Formic acid
Eurylychnus blagravei	Methacrylic acid; tiglic acid
E. olliffi	Methacrylic acid; tiglic acid; isovaleric acid
Harpalus atratus, H. azurus, H. caliginosus, H. dimidiatus, H. distinguendus, H. griseus, H. luteicornis, H. pubescens, H. tardus	Formic acid
Helluo costatus	Formic acid; *n*-nonyl acetate; *n*-nonyl formate
Hellumorphoides ferrugineus, H. latitarsis	Formic acid; *n*-nonyl acetate
Laccopterum foveigerum	Ethacrylic acid; hexenoic acid; methacrylic acid; tiglic acid; *n*-caproic acid; crotonic acid; isocrotonic acid
Carabus auratus	Methacrylic acid; tiglic acid
C. auronitens, C. cansellatus, C. contexus, C. granulatus	Methacrylic acid; tiglic acid
C. problematicus	Methacrylic acid; tiglic acid
C. taedatus	Ethacrylic acid; methacrylic acid
Carenum bonelli	Angelic acid; isocrotonic acid; hexenoic acid; methacrylic acid
C. interruptum	Angelic acid; isocrotonic acid; methacrylic acid
C. tinctillatum	Angelic acid; isocrotonic acid; methacrylic acid; tiglic acid
Castelnaudia superba	Acetic acid; methacrylic acid, tiglic acid
Chlaenius australis, C. hipunctatus, C. chrysocephalus, C. cordicollis, C. festivus, C. tristus	*m*-cresol
C. restitus	*p*-benzoquinone; *p*-ethylquinone; *p*-toluquinone

Table 19 (continued)

Carabidae	Components identified in the secretion
Chrina basalis	*p*-benzoquinone; *p*-toluquinone
C. fossor	*p*-benzoquinone; 2-methoxy-3-methyl-quinone; *p*-toluquinone
Craspedophorus sp.	*m*-cresol; tridecane
Cratoferonia phylarchus, Cratogaster melas	Methacrylic acid; tiglic acid
Lebia chlorocephala	Formic acid
Leistus ferrugineus	Methacrylic acid; tiglic acid
Licenus nitidor	Formic acid
Loricera pilicornis	isobutyric acid; isovaleric acid
Loxandrus longiformis	Salicylaldehyde
Loxadactylus carinulatus	Methacrylic acid
Megacephala australis	Benzaldehyde
Molops elatus	Methacrylic acid; tiglic acid
Mystropomus regularis	*p*-benzoquinone; *p*-ethylquinone; *p*-toluquinone
Nebria livida	Methacrylic acid; tiglic acid
Notiophilus biguttatus	isobutyric acid; isovaleric acid
Notonomus angustibasis, N. crenulatus, N. miles, N. muelleri, N. opulentus, N. rainbowi, N. scotti, N. triplogeniodes, N. variicollis, Odacantha melanura	Formic acid
Omophron limbatum	isobutyric acid; isovaleric acid
Pamborus alternans, P. guerini, P. pradieri, P. viridis	Ethacrylic acid; methacrylic acid
Panagaeus bipistulatus	*m*-cresol
Pasimachus californicus, P. duplicatus, P. elongatus	Methacrylic acid
Paussus favieri	*p*-benzoquinone; *p*-toluquinone
Philophloeus tuberculatus	Crotonic acid; isocrotonic acid; methacrylic acid; 4-methylvaleric acid; tiglic acid
Polystichus connexus, Progaleritina mexicana	Formic acid
Promecoderus spp.	*n*-butyric acid; caproic acid; isovaleric acid

Carabidae	Components identified in the secretion
Prosopognus harpaloides	Methacrylic acid
Pseudoceneus iridescens	Methacrylic acid; tiglic acid
Pterostichus cupreus, P. macer, P. melas, P. metallicus, P. niger, P. vulgaris	Decane; methacrylic acid; tiglic acid; tridecane; undecane
Rhytisternus laevilaterus	Methacrylic acid; tiglic acid
Sarticus cyaneocinctus	Formic acid
Scaphinotus andrewsi germari, S. adrewsi montana, S. viduus, S. webbi	Methacrylic acid; tiglic acid
Siagonyx blackburni, Sphallomorpha colymbetoides, Sphodrosomus saisseti	Formic acid
Stenaptinus catoirei, S. verticalis	*p*-benzoquinone; *p*-toluquinone
Stenolophus mixtus	Formic acid
Thermophilum burchelli, T. homoplatum	Acetic acid; formic acid; tiglic acid; isovaleraldehyde
Trichosternus nudipes	Methacrylic acid; tiglic acid

3.3.3.2 Cerambycidae

The *Cerambycidae* differ in the chemistry of their toxins. Whereas *Stenocentrus ostricilla* and *Sillytus grammicus* both use o-cresol and toluene, *Aromia moschata* uses a mixture of salicylaldehyde and four monoterpenes, namely *cis*-and *trans*-rose-oxide, and γ and δ-iridodial; iridodial has been found in several ant species, too. The secretion of the Australian eucalyptus langicorn beetle, *Phoracantha semipunctata,* contains 2-hydroxy-6-methylbenzaldehyde, phoracanthal and phoracanthol.

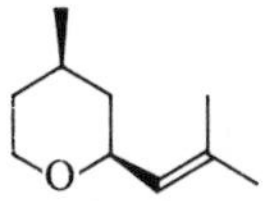

cis-Rose oxide

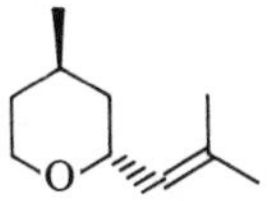

trans-Rose oxide

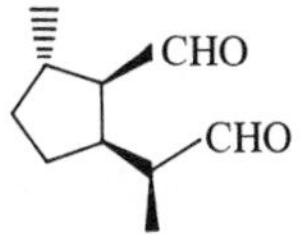

CHO CHO

δ-Iridodial

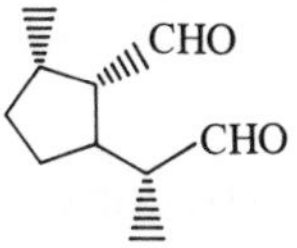

CHO CHO

γ-Iridodial

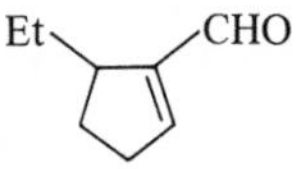

Phoracanthal

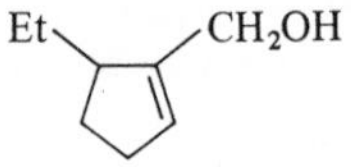

Phoracanthol

3.3.3.3 Coccinellidae

The *Coccinellidae* use alkaloid toxins for their defense; a survey of the distribution is given in Table 20.

Table 20. Distribution of alkaloid toxins in the *Coccinellidae*

Species	Alkaloid						
	Coccinelline	Precoccinelline	Convergine	Hippodamine	Myrrhine	Propyleine	Adaline
Adalia bipunctata							×
A. 10-punctata							×
Anisosticta 19-punctata				×			
Cheilomenes propinqua (var. 4 lineata)	×	×					
Coccinella californica	×						
C. 7-punctata	×	×					
C. 5-punctata	×	×					
C. 11-punctata	×						
C. 14-punctata	×	×					
Coleomegilla maculata		×?			×?		
Hippodamia convergens			×	×			
Micraspis 16-punctata		×					
Myrrha 18-punctata					×		
Propylaea 14-punctata						×	

At present the chemical structure of seven *Coccinellidae* alkaloids is known:

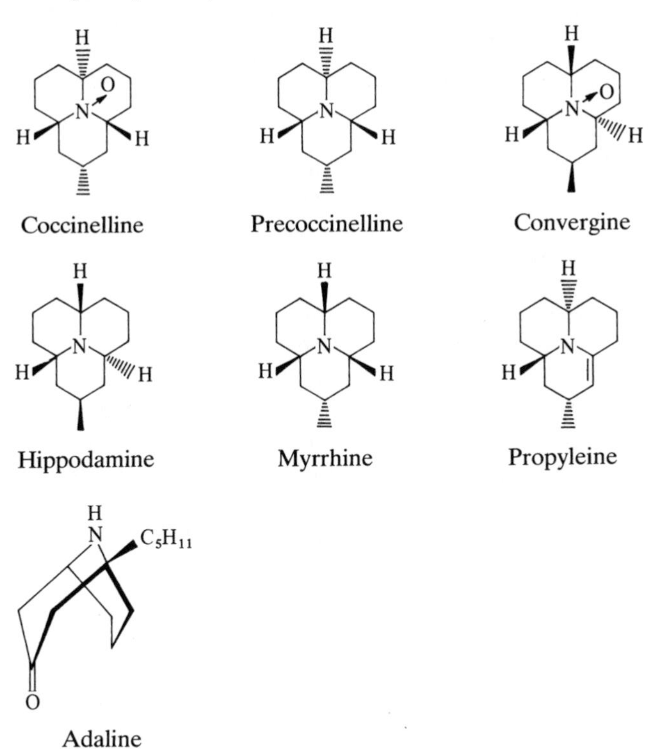

3.3.3.4 Dytiscidae

The *Dytiscidae* (diving beetles) possess in their pygidial glands a mixture of *p*-hydroxy-benzaldehyde, *p*-hydroxy-benzoic acid and benzoic acid, which not only protects against other animals but may

Table 21. Dytiscid toxins

Dytiscidae	Components identified in the secretion
Dytiscus latissimus	Benzoic acid; *p*-hydroxybenzaldehyde; *p*-hydroxybenzoic acid methyl ester
P. marginalis	Benzoic acid; 3,5-dihydroxyphenylacetic acid methyl ester; *p*-hydroxybenzaldehyde; *p*-hydroxybenzoic acid methyl ester; 3-(*p*-hydroxyphenylmethylene)-5-hydroxy-benzo [b]furan-2-one; 4,6-pregnadien-20α-ol-3-one; 4-pregen-21-ol-3,20-dione; 4-pregen-20α-ol-3-one

Table 21 (continued)

Dytiscidae	Components identified in the secretion
Graphoderus cinereus	Benzoic acid; 3,4-dihydroxybenzoic acid methyl ester; *p*-hydroxybenzaldehyde; *p*-hydroxybenzoic acid; *p*-hydroxybenzoic acid methyl ester
Ilybius fenestratus	1,4-androstadien-17β-ol-3-one; 4-androsten-17β-ol-3-one; benzoic acid; 1,3,5-estratrien-3-ol-17-one; 1,3,5estratrien-3, 17β-diol; hydroquinone; *p*-hydroxybenzaldehyde; *p*-hydroxybenzoic acid; *p*-hydroxybenzoic acid methyl ester; 8-hydroxyquinoline carboxylic acid methyl ester; 4-pregnen-20β-ol-3-one
I. fuliginosus	Testosterone
Rhantus exsoletus	Benzoic acid; hydroquinone; *p*-hydroxybenzaldehyde; *p*-hydroxybenzoic acid; *p*-hydroxybenzoic acid methyl ester
Acilius sulcatus	Benzoic acid; *p*-hydroxybenzaldehyde; *p*-hydroxybenzoic acid methyl ester; 4,6-pregnadien-3,20-dione; 4,6-pregnadien-20α-ol-3-one; 4,6-pregnadien-21-ol-3,20-dione; 4-pregnen-20α-ol-3-one; pregnen-21-ol-3,20-dione
Agabus bipustulatus	3,4-dihydroxybenzoic acid methyl ester; *p*-hydroxybenzaldehyde; *p*-hydroxybenzoic acid methyl ester; 4-pregnen-21-ol-3,20-dione
A. sturmi	Benzoic acid; *p*-hydroxybenzaldehyde; *p*-hydroxybenzoic acid methyl ester; 4,6-pregnadien-15α,20β-diol-3-one isobutyrate; 4,6-pregnadien-α-3,20-dione; 4,6-pregnadien-15α-3,20-dione isobutyrate
Colymbetes fuscus	Benzoic acid; 3,4-dihydroxybenzoic acid methyl ester; hydroquinone; *p*-hydroxybenzaldehyde; *p*-hydroxybenzoic acid; *p*-hydroxybenzoic acid methyl ester
C. lateralimarginalis	Benzoic acid; 3,4-dihydroxybenzoic acid ethyl ester; 3,4-dihydroxybenzoic acid methyl ester; *p*-hydroxybenzoic acid methyl ester; 4,6-pregnadien-12β,20α-diol-3-one; 4,6-pregnadien-3,20-dione; 4,6-pregnadien-20α-ol-3-one; 4,6-pregnadien-21-ol-3,20-dione
C. limatus	4,6-pregnadien-20α-ol-3-one; 4,6-pregnadien-21-ol-3,20-dione; 4,6-pregnadien-12β-ol-3,20-dione; 4-pregnen-20α-ol-3-one; 4-pregnen-21-ol-3,20-dione; 4-pregnen-12β-ol-3,20-dione
C. tripunctatus	Benzoic acid; *p*-hydroxybenzaldehyde; *p*-hydroxybenzoic acid methyl ester; 4,6-pregnadien-12β-ol-3,20-dione; 4,6-pregnadien-20α-ol-3-one; 4-pregnen-20β-ol-3-one; 4-pregnen-21-ol-3,20-dione

also be used to protect against microorganisms during hibernation in the mud of ponds. It is remarkable and surprising that in the prothoriac glands steroids are found that are otherwise found as hormones in vertebrates. *Dytiscus marginalis,* for example, possesses relatively large amounts (0.4 mg) of cortexon. Such quantities disturb the K^+/Na^+ relation of organisms, and fish lose "consciousness" shortly after eating one of these beetles. They soon learn that those beetles are inedible.

The distribution of the Dytiscid toxins is presented in Table 21.

3.3.3.5 Gyrinidae

The pygidial glands of the *Gyrinidae* contain norsesquiterpenoids:

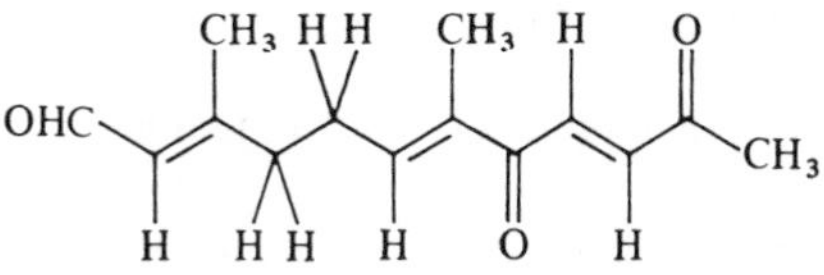

Gyrinidal

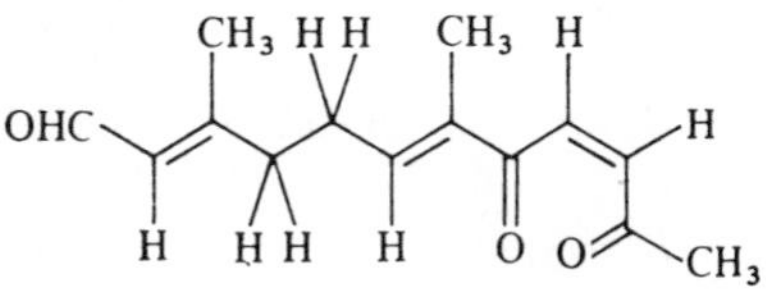

Isogyrinidal

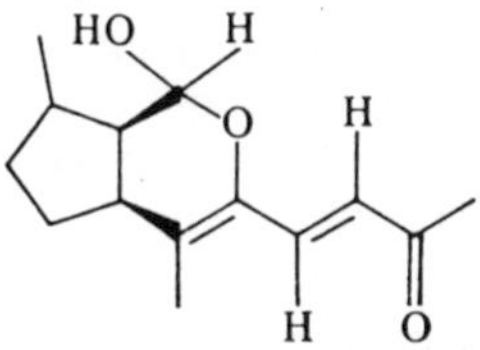

Gyrinidone

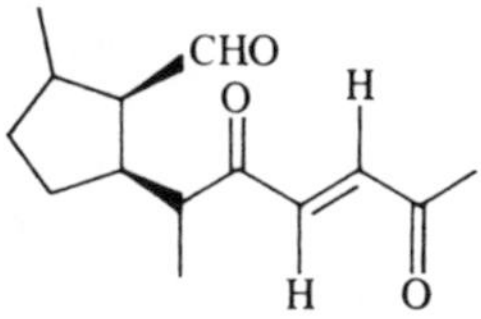

Gyrinidione

3.3.3.6 Meloidae

The toxin of the blister beetles has received considerable attention since medieval times: It is the supposed aphrodisiac known as "Spanish fly". The structure of this toxin, Cantharidin, is a follows:

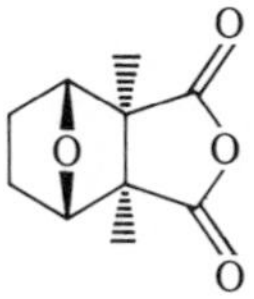

Cantharidin

3.3.3.7 Silphidae

The carrion beetles *Oecoptoma thorica, Phospuga etrata* and *Silpha obscura* use ammonia as a defense agent.

3.3.3.8 Staphylinidae

The *Staphylinidae* toxins are most diverse. They are listed in the following Table 22:

Table 22. The components of Staphylinid toxins

Staphylinidae	Components identified in the secretion
Bledius mandibularis	γ-dodecalactone; geranial; neral; *p*-toluquinone; 1-undecene
B. spectablis	γ-dodecalactone; geranial; methyl benzoquinone; neral; 1-undecene
Creophilus erythro-cephalus	Iridodial
Drusilla canaliculata	dodecanal; hydroquinone; 2-hydroxy-3-methylhy-droquinone; 2-methoxy-3-methylhydroquinone; 3-methoxy-*p*-toluquinone; 2-methylhydroquinone; tetradec-5-enal; tetradeca-5,8-dienal; tetradecanal; *p*-toluquinone; tridec-4,7-diene; tridec-4-ene; tridecane; undecane
Eulissus orthodoxus	Citronellal; iridodial; isovaleraldehyde
Hesperus semirufus	Actinidine
Lomechusa strumosa	*p*-benzoquinone; *p*-ethylquinone; *p*-toluquinone; tridecane; 1-tridecene
Paedenus fuscipes	Pederin, pederone and pseudopederin
Philonthus politus	Actinidine
Staphylinus olens	Iridodial; 4-methylhexan-3-one
Stenus bipunctatus	1,8-cineole; 6-methylhept-5-en-2-one; isopiperitenol
S. comma	1,8-cineole; N-ethyl-3-(2-methylbutyl)-piperidine; 6-methylhept-5-en-2-one; isopiperitenol
Thyreocephalus lorquini	Citronellal; iridodial; isovaleraldehyde
Zyras humeralis	isovaleric acid

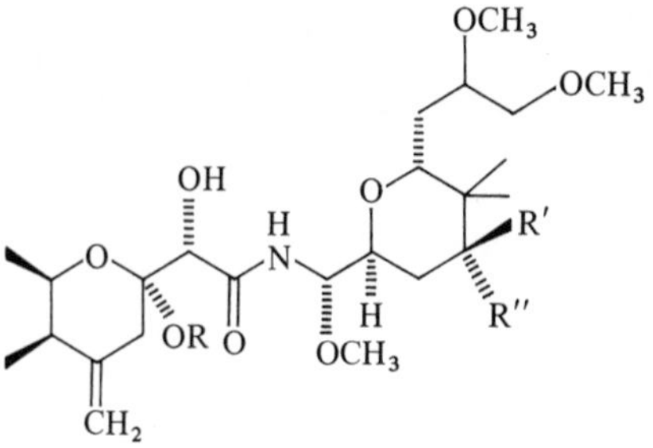

Pederin	R = CH_3; R′ = H; R″ = OH
Pseudopederin	R = H; R′ = H; R″ = OH
Pederone	R = CH_3; R′+R″ = O

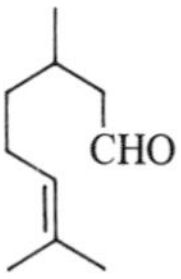

Citronellal 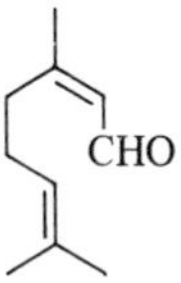 Neral 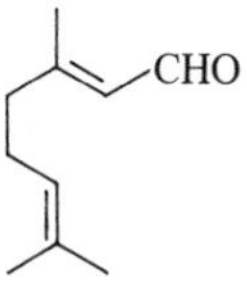 Geranial 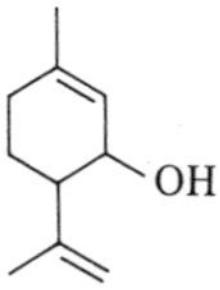 Isopiperitenol

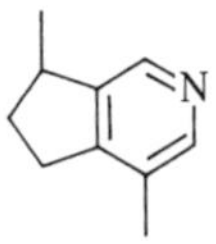

Actinidine 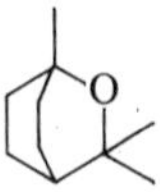 1,8-Cineole 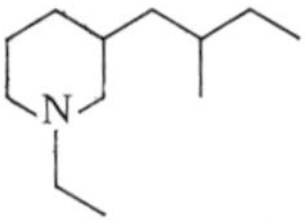 N-Ethyl-3-(2-methylbutyl)-piperidine

Table 23. The distribution of defensive secretions within the *Tenebrionidae*

Tenebrionidae	Components identified in the secretion
Alphitobius diaperinus	*p*-ethylquinone; *p*-methylquinone
Amphidora littoralis	*p*-ethylquinone; 2-methoxy-3-methylquinone; *p*-toluqinone; 1-tridecene
A. nigropylosa	*p*-ethylquinone; 2-methoxy-3-methylquinone; 1-pentadecene; *p*-toluquinone; 1-tridecene
Argoporis alucatecea,	*p*-benzoquinone; 6-butyl-1,4-naphtoquinone;

Table 23 (continued)

Tenebrionidae	Components identified in the secretion
A. rufipes	6-ethyl-1,4-naphtoquinone; *p*-ethylquinone; 6-methyl-1,4-naphtoquinone; 6-propyl-1,4-naphtoquinone; *p*-toluquinone
Blaps approximans, B. juliae, B. lethifera	*p*-benzoquinone; *p*-ethylquinone; *p*-toluquinone; 1-tridecene
B. mortisaga, B. mucronata, B. requienii, Diaperis boleti, D. maculata	*p*-ethylquinone; *p*-toluquinone
B. polycresta	*p*-benzoquinone; *p*-ethylquinone; *p*-toluquinone
B. proheta, B. stranchi	*p*-ethylquinone; *p*-toluquinone; 1-tridecene
B. sulcata, B. wiedermanni	*p*-benzoquinone; 2-ethylhydroquinone; *p*-ethylquinone; hydroquinone; 2-methylhydroquinone; *p*-toluquinone; 1-tridecene
Bolitotherus cornutus	*p*-ethylquinone; *p*-toluquinone
Cratidus osculans	*p*-ethylquinone; 1-heptadecene; 1-nonene; *p*-toluquinone; 1-tridecene; 1-undecene
Eleates occidentalis	*p*-ethylquinone; 1-heptadecene; *p*-toluquinone
Eleodes acuta, E. armata	*p*-ethylquinone; 1-nonene; *p*-toluquinone; 1-tridecene; 1-undecene
E. acuticauda	*p*-benzoquinone; *p*-ethylquinone; 1-nonene; octanoic acid; *p*-toluquinone; 1-tridecene; 1-undecene
E. aristatus	*p*-ethylquinone; 2-methyl-3-methoxybenzoquinone; 1-pentadecene; *p*-toluquinone; 1-tridecene
E. blanchardi	*p*-ethylquinone; 2-methoxy-3-methylbenzoquinone; 1-nonene; *p*-toluquinone; 1-tridecene; 1-undecene
E. carbonaria	*p*-ethylquinone; 1-pentadecene; *p*-toluquinone; 1-tridecene; 1-undecene
E. caudifera	*p*-benzoquinone; *p*-ethylquinone; 1-heptadecene; *p*-toluquinone; 1-tridecene; 1-undecene
E. consobrinus, dentipes, E. gigantea, E. nigrina	*p*-ethylquinone; 1-nonene; *p*-toluquinone; 1-tridecene; 1-undecene
E. constrictus, E. incultis	*p*-ethylquinone; 2-methoxy-3-methylbenzoquinone; *p*-toluquinone; 1-tridecene
E. cordatus	*p*-ethylquinone; 2-methoxy-3-methylbenzoquinone; 1-nonene; *p*-toluquinone; 1-tridecene; 1-undecene

Tenebrionidae	Components identified in the secretion
E. dissimilis	*p*-benzoquinone; *p*-ethylquinone; 1-nonene; octanoic acid; *p*-toluquinone, 1-tridecene; 1-undecene
E. extricata	*p*-benzoquinone; *p*-ethylquinone; 1-nonene; *p*-toluquinone; 1-tridecene; 1-undecene
E. femorata	*p*-ethylquinone; octanoic acid; *p*-toluquinone; 1-tridecene; 1-undecene
E. goryi	*p*-ethylquinone; 1-heptadecene; 1-pentadecene; *p*-toluquinone; 1-tridecene; 1-undecene
E. gracilis, E. hispilabris	*p*-ethylquinone; 1-nonene; octanoic acid; *p*-toluquinone; 1-tridecene; 1-undecene
E. grandicollis	*p*-ethylquinone; 2-methoxy-3-methylbenzoquinone; 1-nonene; octanoic acid; *p*-toluquinone; 1-tridecene; 1-undecene
E. knullorum	*p*-ethylquinone; 1-nonene; 1-pentadecene; *p*-toluquinone; 1-tridecene; 1-undecene
E. laticollis	*p*-ethylquinone; 1-nonene; octanoic acid; *p*-toluquinone; 1-tridecene; 1-undecene
E. longicollis	*p*-benzoquinone; *p*-ethylquinone; *p*-methylquinone; octanoic acid; 1-tridecene; 1-undecene
E. longipilosa	*p*-ethylquinone; *p*-toluquinone; 1-tridecene
E. neotomae	*p*-ethylquinone; 2-methoxy-3-methylbenzoquinone; 1-nonene; *p*-toluquinone; 1-tridecene; 1-undecene
E. obsoleta	*p*-benzoquinone; *p*-ethylquinone; 2-methoxy-2-methylbenzoquinone; 1-nonene; 1-pentadecene; *p*-toluquinone; 1-tridecene; 1-undecene
E. omissa, E. parowana	*p*-ethylquinone; *p*-toluquinone; 1-tridecene; 1-undecene
E. pimelioides	*p*-ethylquinone; 2-methoxy-3-methylbenzoquinone; 1-nonene; *p*-toluquinone; 1-tridecene
E. sponsa	*p*-benzoquinone; *p*-ethylquinone; 1-nonene; octanoic acid; *p*-toluquinone; 1-tridecene; 1-undecene
E. subnitens	*p*-ethylquinone; 1-pentadecene; *p*-toluquinone; 1-tridecene; 1-undecene
E. tenebrosus	*p*-ethylquinone; 2-methoxy-3-methylbenzoquinone; *p*-toluquinone; 1-tridecene; 1-undecene
E. tenuipes	*p*-ethylquinone; 1-nonene; *p*-toluquinone; 1-tridecene; 1-undecene

Table 23 (continued)

Tenebrionidae	Components identified in the secretion
E. tricostata	*p*-benzoquinone; *p*-ethylquinone; *p*-toluquinone; 1-undecene
E. obscura	*p*-ethylquinone; 1-heptadecene; 1-nonene; octanoic acid; *p*-toluquinone; 1-tridecene; 1-undecene
E. ventricosa	*p*-ethylquinone; 1-nonene; 1-pentadecene; *p*-toluquinone; 1-tridecene; 1-undecene
Embaphrion muricatum	*p*-benzoquinone; caprylic acid; *p*-ethylquinone; 1-nonene; *p*-toluquinone; 1-tridecene; 1-undecene
Epanitus obscurus, Gnaptor spinimanus, Opatroides punctulatus	*p*-ethylquinone; *p*-toluquinone
Eurynotus capensis	*p*-ethylquinone; 1-heptadecene; *p*-toluquinone; 1-tridecene
Gnathocerus cornutus, Gonocephalum arenarium, Gonopus agrestis, P. oregonense, P. subcostatum	*p*-ethylquinone; *p*-toluquinone; 1-tridecene
G. kolbi	*p*-ethylquinone; *p*-toluquinone; 1-tridecene; 1-undecene
G. tibialis	*p*-ethylquinone; 1-heptadecene; 1-nonene; 1-pentadecene; *p*-toluquinone; 1-undecene
Helops aenus, H. quisquilus	*p*-ethylquinone; *p*-toluquinone
Iphthimus laevissimus, I. serratus	*p*-benzoquinone; *p*-ethylquinone; *p*-toluquinone
Latheticus oryzae	2-ethylquinone; 2-methylquinone
Leichenum canaliculatum variegatum, Pyanisia tristis, S. uncinus	*p*-benzoquinone; *p*-ethylquinone; *p*-toluquinone
Melanopterus marginicollis	*p*-ethylquinone; 1-nonene; 1-pentadecene; *p*-toluquinone; 1-undecene
M. porcus	1-nonene; *p*-toluquinone; 1-undecene
Mercantha contracta	*p*-benzoquinone; *p*-ethylquinone; 1-heptadecene; 1-nonene; 1-pentadecene; *p*-toluquinone
Parastizopus balneorum	*p*-ethylquinone; 1-heptadecene; 1-nonene; 1-pentadecene; *p*-toluquinone; 1-undecene

Tenebrionidae	Components identified in the secretion
Phaleria rotundata, P. testacea	*p*-ethylquinone; 1-heptadecene; 1-pentadecene; *p*-toluquinone
Platydema americanum	*p*-ethylquinone; *p*-toluquinone; 1-tridecene; 1-undecene
P. flavipes, P. ruficorne	*p*-ethylquinone; 1-pentadecene; *p*-toluquinone; 1-tridecene
Psorodes calcaratus, Schelodontes sp.	*p*-ethylquinone; 1-nonene; 1-pentadecene; *p*-toluquinone; 1-undecene
P. gratilla	*p*-ethylquinone; 1-heptadecene; 1-nonene; *p*-toluquinone; 1-undecene
Scaurus aegytiacus	*p*-ethylquinone; *p*-toluquinone; 1-tridecene; 1-undecene
Tenebrio molitor, Pimelia confusa, Morisia planta tingitiana	*p*-toluquinone
T. obscurus	*p*-benzoquinone
Tribolium brevicornis	*p*-ethylquinone; 1-pentadecene; *p*-toluquinone
T. castaneum	*p*-ethylquinone; 2-hydroxy-4-methoxy-propiophenone; *p*-methoxyquinone; *p*-toluquinone
T. confusum	*p*-ethylquinone; 2-hydroxy-4-methoxy-propiophenone; 1-pentadecene; *p*-toluquinone
T. destructor	*p*-ethylquinone; 1-pentadecene; *p*-toluquinone
Trigonopus capicola	*p*-ethylquinone; *p*-toluquinone
Zadenos delandei, Z. longipalpus	*p*-ethylquinone; 1-heptadecene; 1-nonene; *p*-toluquinone; 1-tridecene
Z. mulsanti	*p*-ethylquinone; 1-heptadecene; 1-nonene; *p*-toluquinone; 1-tridecene; 1-undecene
Zophobas rugipes	*p*-benzoquinone; *m*-cresol; *m*-ethylphenol; *p*-ethylquinone; phenol; *p*-toluquinone

References

Schildknecht, H., Siewerdt, R., Maschwitz, U.: Angew. Chem. *78*, 392 (1966)

Schildknecht, H. et al.: 4-Pregnen-15α, 20β-diol-3-on im Wehrsekret eines Schwimmkäfers. Naturwissenschaften *56*, 37 (1969)

Schildknecht, H. et al.: Explosionschemie der Bombardierkäfer. Naturwissenschaften *56*, 328 (1969)

Schildknecht, H. et al.: Blausäure im Wehrsekret des Erdläufers. Naturwissenschaften *55*, 230 (1968)

Schildknecht, H. et al.: Zur Evolution der Carabiden-Wehrdrüsensekrete. Naturwissenschaften *55*, 112 (1968)

Diplopoda

Eisner, T., Alsop, D., Hicks, K., Meinwald, J.: In "Arthropod Venoms" (Ed. S. Bettini) p. 45–47, 52 (1978)

Hemiptera

Weatherstone, J., Percy, J. E.: In "Arthropod Venoms" (Ed. S. Bettini) p. 496–501 (1978)

Coleoptera

Weatherstone, J., Percy, J. E.: In "Arthropod Venoms" (Ed. S. Bettini) p. 513–517, 520–542 (1978)

Myriapoda

Minelli, A.: In "Arthropod Venoms" (Ed. S. Bettini), p. 81 (1978)

3.3.4 Lepidoptera

Lepidoptera, adult moths, larval forms and sometimes also pupae may affect human health. The weapons utilized by these animals are specialized scales or toxic substances or both; they are used actively in predation as well in defense against predators.

Two groups can be differentiated. One is called *cryptotoxic endogenous lepidoptera.* They are toxiferous but have no specialized apparatus for secreting their venoms. Their toxicity is related to food habits or other ecologic factors. For example they can affect cattle that have accidentally eaten larvae or pupae. Many fatalities in cattle have been observed, especially in South America. Indigenous humans e. g. the pygmies in the Kalahari desert, South West Africa, use pupae for the preparations of dart poisons.

This group is discussed in Chapter 4.3.6.

The second group are the *phanerotoxic exogenous lepidoptera.* The anthropotoxic erucae (lat. eruca = caterpillar) possess special glands that eject the venoms. Erucism is observed all over the world, especially on the American continents. Cases have been reported from the Old World since ancient times, but no statistics are available on number or cause of accidents, except for one report of 433 cases in Peru 1958 and 1962.

Poisoning

The local reactions are caused by contact with the poison hairs. The skin or mucous membranes display inflammation, redness, swelling and pain. In severe cases hemorrhage and necrosis are observed. Besides the local symptoms general symptoms may occur, such as fever, nausea and vomiting, general malaise, spasms and paralysis. The general symptoms are obviously produced by peptide substances, since a desensitation can be observed; this kind of immunity, however, is temporary.

The symptoms disappear between 24 hours and 7 days without treatment, but secondary infections have to be taken into account, and in some cases lesions have developed. Otherwise the patients have to be treated symptomatically.

Lepidopterism is the manifestation of poisoning caused by adult moths. Lesions and necrosis are the local reactions, the general symptoms being the same as above.

Lepidopterism is also frequent in South America. In epidemic outbreaks reported in Brazil and Peru, locally 40–80% of the population suffered from lepidopteran dermatitis.

3.3.5 Hymenoptera: Aculeata

3.3.5.1 Apidae (Bees) and Vespidae (Wasps)

Bees, wasps and hornets have so many things in common with respect to their venoms that they can be discussed together. A highly specialized venom apparatus consisting of gland, canal and sting provided for the production, storage and ejection of the venom; it is used mainly for defense. In contrast to other animal venoms, these consist to only a minor degree of enzymes; the main components are biogenic amines, polypeptides and kinins. (Kinins: biologic active peptides acting on the smooth musculature and possessing hypotensive and hypertensive activities.) A review is given in Table 24.

Table 24. Compounds of the venom from apoids and vespoids

	Bees	Wasps	Hornets
Biogenic amines	Histamine	Histamine Serotonin	Histamine Serotonin Acetycholine
Peptides	Apamin Melittin Mast-cell degranulating peptide	Wasp-kinin	Hornet-kinin
Enzymes	Phospholipase A Hyaluronidase	Phospholipase A Phospholipase B Hyaluronidase	Phospholipase A Phospholipase B Hyaluronidase

Poisoning

The local reactions of bee, wasp or hornet stings are well known. The painful swellings are caused by all components contained in the venom; the symptoms overlap each other. The pain is mainly due to histamine and serotonin.

The toxicity of wasp and hornet venom is not exactly known; the LD_{50} of purified lyophilized venom is 2.5 mg/kg (mouse, i. v.). The LD_{50} of bee venom is 6 mg/kg (mouse, i. v.). Per sting about 0.1 mg venom is injected into the victim; thus more than 100 stings would be necessary to kill a human.

On the other hand fatal stings by bees, wasps or hornets are not at all rare (cf. Table 1, p. 5). We know of many cases in which one or two stings killed a person, but there are also some rare cases in which persons attacked by the Brazilian killer bees survived 400 or more stings. The cause of death therefore cannot be attributed to the venom alone, but to anaphylactic reactions. Practically all cases show the signs of an anaphylactic reaction. Allergic persons are especially sensitive. The other allergic phenomena are itching urticaria, allergic asthma and in severe cases circulatory failure. These general symptoms are usually observed shortly after the sting and disappear within several hours. In fatal cases death happens within one hour or less.

Fever (39.4 °C), hypotension (85/65) and tachycardia (pulse rate of 120) may occur. Cardiac and respiratory arrest have been observed, too.

Treatment

In most cases, single stings do not produce serious symptoms except for anaphylaxia, which is limited to few victims. In case of multiple stings, hospitalization is indicated in order to cope with the immediate dangers as well as delayed reactions that might occur even after 24–48 hours. Parenteral infusions of antihistamines and of hydrocortisone are administered. Special attention should be given to the danger of renal failure or of acute hemolysis, which may require blood transfusion. There are no specific antivenoms.

The treatment of single sting cases can be limited to non-specific analgesics. In allergic cases the usual therapy with calcium, adrenaline and corticosteroids is indicated.

The antigenic properties of hymenoptera venoms are of theoretical as well as of practical interest. Injection of crude venom raises the antiphospholipase titre and the antihyaluronidase titre in rabbits and guinea pigs. This is to be expected, since phospholipase and hyaluronidase are typical proteins. The desensitization and immunity of bee keepers may be due to this fact. The very low antigen properties of melittin and apamin can be attributed to the low molecular weight; neither compound can be included among the proteins, since they possesss 26 and 18 amino acids only. Melittin, however, is responsible for many pharmacologic properties of the venoms.

Chemistry

A. Biogenic Amines

Histamine, serotonin and acetylcholine are pain producing substances, and they probably are responsible for the pain that occurs immediately after the sting. Serotonin is responsible for the acute effect on the blood vascular system, whereas the cardiac activity of hornet venom stems almost exclusively from acetylcholine. The three substances have been identified beyond any doubt. However, their relative quantities are still debatable: The values for histamine and serotonin vary between 1 and 3% of dry weight; the contents of acetylcholine in hornet venom is about 5% of dry weight.

B. Peptides

Kinins

Kinins are substances that influence blood pressure (mostly hypotensively) and act on the smooth musculature; the rapid reversal of effects is characteristic. Bee venom is free of kinins, but they are found in wasp and hornet venom. Wasp kinin is difficult to dialyse and very stable against heat in moderate acid or neutral solutions. It is extremely hypotensive and raises the vascular permeability of the skin. Moreover is produces pain. By paper chromatography it can be easily differentiated from plasmakinin, bradykinin and kallidin. Hornet kinin is similar but not identical to wasp kinin. The qualitative activity is the same; hornet kinin, however, is somewhat weaker in activity. The importance of the kinins for the whole composition of the crude venom is hard to estimate, as there are no reliable and unambigous data on their concentration in the crude venom.

Apamin

Apamin makes up 2% of the dry bee venom. Paper electrophoresis in phospate buffer of pH 7.0 permits easy separation from the other constituents. Refinement can be achieved by chromatography on carboxymethyl cellulose. Apamin is a peptide with 18 amino acids; the sequence is as follows:

Cys-Asn-Cys-Lys-Ala-Pro-Glu-Thr-Ala-Leu-Cys-Ala-Arg-Arg-Cys-Glu-Glu-His-NH_2

The pharmacologic activity of apamin on the central nervous system is noteworthy. Blood pressure, however, is not influenced. Sublethal doses produce a hypermotility that may last for some days. The LD_{50} is 4 mg/kg (mouse, i. v.). Considering the whole activity of bee venom, apamin plays a minor role because of its lower concentration.

Melittin

Melittin is the main constituent of bee venom; it amounts to 50% of dry substances. It can be separated from phospholipase A and from hyaluronidase by precipitation with picric acid, since it is less soluble. Refinement is done first by electrophoresis. The last traces of apamin, which may be carried on, can be removed by gel filtration on Sephadex G-50 and subsequent chromatography on carboxymethyl cellulose. The amino acid sequence is as follows:

Gly-Ile-Gly-Ala-Val-Leu-Lys-Val-Leu-Thr-Thr-Gly-Leu-Pro-Ala-Leu-Ile-Ser-Trp-Ile-Lys-Arg-Lys-Arg-Gln-Glu-NH_2

It is remarkable that neither sulphur-containing amino acids nor phenylalanin, tyrosine or histidine are present in this sequence. Melittin can be cleaved by pepsin, trypsin and chymotrypsin. Melittin is extremely basic; the isoelectric point is at pH 10. This causes the surface tension of water to decrease, and to a degree not known with any other peptide. This basicity may also explain the "direct" hemolytic activity. The amino acid sequence shows a structure similar to an invert soap: Position 1–20 are occupied by neutral, hydrophobic residues; the C-terminal part consists of basic components.

Pharmacology of Mellitin: The hemolytic activity of melittin has been mentioned. Blood plasma, lecithin, polysaccharide sulfate and in higher concentrations citrate lower the hemolytic activity of mellitin, probably by acid/base reaction. The hemolytic activity in reality is not so strong as to be lethal. During and immediately before hemolysis potassium is set free. This is a special case of the liberation of a pharmacologically active substance by melittin. Damage of other potassium-barriers could be responsible for additional effects. Mast cells, which contain histamine and serotonin, are damaged as well as erythrocytes, and in the same manner serotonin is set free from thrombocytes.

The contraction of the smooth musculature is not only due to the liberation of these biogenic amines. Melittin contraction characteristically differs from that of histamine by a slower onset and a longer duration. Atropine acts antagonistically at low doses, papaverin at higher.

Melittin has both stimulatory and paralytic effects on heart function. Low doses possess a positive inotropic effect, higher doses lead to an irreversible contraction. The blood pressure falls to very low values for a short time. Repeated application of mellitin, however, results in small decreases only, which finally change to a longer lasting increase. These effects result from a direct action of mellitin on heart and circulation; it is still uncertain if liberation of potassium is important. The LD_{50} is 3.5 mg/kg (mouse, i. v.). There are no characteristic symptoms accompanying lethal doses. The cause of death cannot yet be explained in detail.

Mast Cell Degranulating Peptide

The role of histamine in bee sting poisoning has been mentioned above. The relatively small amount is insufficient to explain the physiologic effects of the whole venom. Melittin and phospolipase A are capable of releasing histamine by destruction of mast cells. But it has turned out that a third peptide, the MCD peptide, is much more powerful, although it is a minor component of the venom (about 2%). The LD_{50} is 40 mg/kg (mouse, i. v.). The structure has been determined as follows:

```
 ┌──────────S       S──────────┐
 │          |       |          │
 │ Ile -Lys-Cys-Asn-Cys-Lys    │
 │                       |     │
 │ Pro-Lys-Ile -Val -His-Arg   S
 │  |                          |
 │ His-Ile -Cys-Arg-Lys-Ile -Cys-Gly
 │           |                    |
 └───────────S                   Lys
                                  |
                                 Asn(NH2)
```

Phospholipase A

The structure of phospholipase A was shown to be:

```
H2N-Ile -Ile -Tyr-Pro-Gly-Thr-Leu-Trp-Cys-Gly-His -Gly-Asn
                                                         |
Thr -His-Lys-Phe-Arg-Gly-Leu-Glu-Asn-Pro-Gly-Ser -Ser -Lys
 |
Asp -Ala-Cys-Cys-Arg-Thr-His -Asp-Met-Cys-Pro-Asn-Val -Met
                                                         |
Ser -Ala-Thr-Asp-Thr-Leu-Gly-His -Lys -Ser -Glu-Gly-Ala-Ser
 |
Arg -Leu-Ser -Cys-Asn-Asp-Asn-Asp-Leu-Phe-Tyr-Lys-Asp-Ser
                                                         |
Phe -Tyr-Met-Lys-Gly-Val -Phe-Tyr -Ser -Ser -Ile -Thr-Asp-Ala
 |
Asn -Leu-Ile -Asn-Thr-Lys-Cys-Tyr-Lys-Leu-Glu-His-Pro-Val
                                                         |
Tyr -His-Leu-Cys-Arg-Gly-Glu-Thr-Arg-Glu-Gly-Cys-Gly-Thr
 |
Thr -Val-Asp-Lys-Ser -Lys-Pro-Lys-Val -Tyr-Gln-Trp-Phe-Asp
                                                         |
                                          Tyr-Lys-Arg-Leu
```

References

Maschwitz, U. W., Kloft, W.: In "Venomous Animals and Their Venoms:, Vol. III, chapter 44 (W. Bücherl and E. E. Buckley, Eds.) New York: Academic Press 1971
Habermann, E.: ibid., chapter 45
Jentsch, J.: Chemie in unserer Zeit *8*, 177 (1974)
O'Connor, R., Peck, M. L.: In "Arthropod Venoms" (Ed. S. Bettini) p. 628, 635 Springer, Berlin, Heidelberg (1978)

3.3.5.2 Formicinae (Ants)

The superfamily *Ants* comprises about 260 genera with 6000 species. 200 species are native to Europe; 80% of all species live in tropical countries.

Most ants except the *Formicinae* are capable of stinging, and in many cases the reactions are severe enough to require medical care. Substances from the Dufour glands as well as from venom glands can be discharged through the sting. In addition compounds generated in the mandibular glands can be used for defense. To what extent these substances my be considered true toxins is debatable in many cases, since at least some of them are used as pheromone or as marking secretions. According to their morphology, the ants are classified into 9 families, seven of which will be discussed here:

a) *Myrmeciinae* (Bulldog Ants)
These ants are well known for their stinging propensities. They occur in Australia.

b) *Ponerinae* (Sting Ants)
They possess a well developed sting. All species are carnivores that feed upon insects. They are found in tropical countries.

c) *Dorylinae* (Migratory Ants)
They are migratory, very aggressive ants whose raids spare no creature. They possess a kind of sting; their biotope is the tropics and subtropics.

d) *Pseudomyrmecinae*
The *Pseudomyrmecinae* possess a primitive venom apparatus.

e) *Myrmicinae*
They possess a well developed venom apparatus; the sting is in some species retarded as for example with *Atta sexdens.* Those species use

their mandibles for defense. They are widely distributed from the tropics to the temperate zones.

f) *Formicinae* (Scaly Ants)

They possess no sting, but the venom gland from which the venom is sprayed still exists. They are not restricted to a specific climate zone; rather they are found all over the world.

g) *Dolichoderinae* (Gland Ants)

They have a retarded sting. The defense function has been taken over by the anal glands.

Envenomation and Treatment

Envenomation occurs either by stinging or spraying. Only a few genera are involved in accidents with humans.

Ponerinae: Probably the most painful stings are delivered from *Paraponera clavata,* native to America. They are described as fierce and extremely aggressive. Intractable pain lasts for many hours. Paralyzing symptoms are produced, and finally a large blister appears. Body temperature may also rise.

Myrmecinae: Extremely painful stings are caused by *Pogonomyrmex* ants. Edematous and erythematous areas occur locally. Children that have been stung sometimes suffer from nausea and vomiting.

In recent years the venom of the fire ant *Solenopsis* has been under intense investigation since they are posing an increasing threat to humans in the southeastern states of the USA, especially in Alabama, Georgia and Mississippi, Texas and northern Florida. Each year about 10,000 stings occur, and the US government has spent several millions of dollars in vain to get rid of this problem. Eight species are responsible for envenomations. The stings of three species, originally indigenous to the US, *S. geminata, S. xyloni* and *S. aurea* are relatively harmless; however, *S. invicta* and *S. richteri,* are feared for their aggressiveness and also because of the severity of the symptoms. Both species were introduced through the harbor of Mobile, Alabama, in the beginning of this century. They flourished in the new environment and today are among the dominant insects of this region. *S. saevissima* and *S. bondari* were also imported from South America. In South Africa *S. punctaticeps* is endemic. The local reaction varies between a slight inflammation or irritation of the skin and severe urticaria, swelling and prurience. In any case the stings are very painful. Pustules may form within 24 hours, which usually turn into necrosis with open wounds. Systemic symptoms are wheezing, shortness of breath and nausea. In some cases unconsciousness and death have been reported, but usually as a result of multiple stings and ensuing anaphylactic shock. There is no specific antivenom available, and treatment is directed toward the symptoms.

Formicinae: The toxin longest known from ants is formic acid. This acid is the protective toxin of the stingless *Formicinae.* One ant may possess according to size and species between 0.005 mg *(Plagiolepis pygmala)* and 4.6 mg *(Camponatus ligniperda).* This may amount to 0.5 to 20% of body weight, the highest values being found with the venom spraying European *Formica rufa.* The high concentration of formic acid within the secretion is astonishing: 70%. The action of formic acid is directed primarily against other insects, where it acts as a respiratory venom.

Chemistry of Venoms and Other Defense Secretions

As mentioned above, defense substances in ants are produced not only in the venom glands but also in the mandibular and abdominal glands (Dufour's glands). They are directed primarily against other small animals such as beetles.

Myrmeciinae. The venom glands of *Myrmecia spp.* contain proteins having a molecular weight between 11000 and 23000, histamine and enzymes such as hyaluronidase, phospholipase A and histamine releasing factors. The Dufour glands produce alkanes and alkenes, e. g. cis-8-heptadecene (62%), n-pentadecane (17%), n-tetradecane (1%), heptadecane (4%) and hexadecane (1%). The remaining 15% are branched hydrocarbons.

Myrmicinae. In *Solenopsis spp.* piperidine alkaloids of the following type have been found:

H_3C–(piperidine, NH)–$CH_2(CH_2)_n \cdot CH_3$

n = 9, 11, 13
cis- or *trans-*2-methyl-6n-alkylpiperidine

H_3C–(piperidine, NH)–$CH_2(CH_2)_2CH{=}CH(CH_2)_7 \cdot CH_3$

Alkaloids with unsaturated side chain

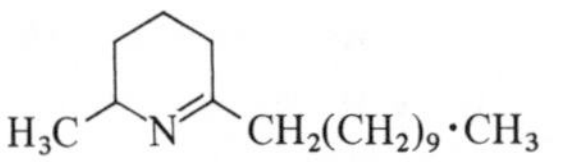

H_3C N $CH_2(CH_2)_9 \cdot CH_3$ Δ^1-Piperidein-derivatives

They are responsible not only for the symptoms in human envenomation but also act as strong insecticides. The South African species *S. punctaticeps* contain pyrrolidine rather than piperidine alkaloids:

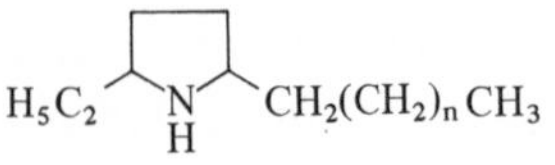

H_5C_2 N H $CH_2(CH_2)_n CH_3$ n = 9, 11, 13
Dialkyl-pyrrolidine

Ponerinae. No investigations of the venoms of ponerine ants have been done, but preliminary analyses have shown that proteins may be responsible for the various activities.

Dorylinae: 4-methyl-3-heptanone in *Neivamyrmex spp.* as well as 3-methylindole

Pseudomyrmex: 4-methyl-3-heptanone, 3-methyl-2-heptanone, 5-methyl-2-heptanone, 6-methyl-3-heptanone, 2-octanone, 2-nonanone

Myrmicinae: C_7-C_{10}-ketones, citronellol, geraniol, neral, geranial, benzaldehyde

Formicinae: Besides the terpenes and ketones mentioned so far from *Dendrolasius fuliginosus,* a modified sesquiterpene, dendrolasin, has been isolated:

Dendrolasin O CH_3 CH_3 CH_3

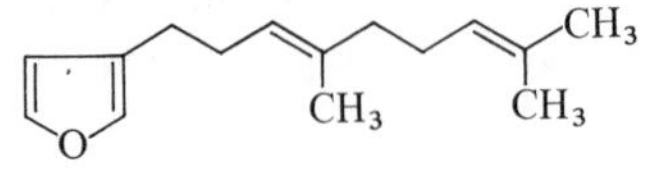

Dolichoderinae: From Iridomyrmex spp. have been isolated:

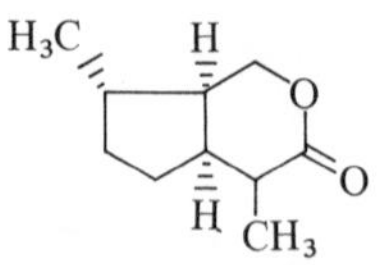

H_3C H O O H CH_3

Iridomyrmecin

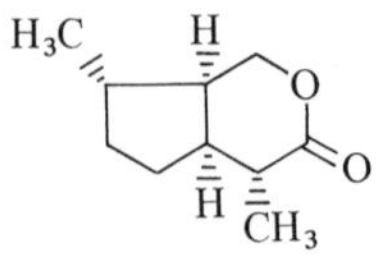

H_3C H O O H CH_3

Isoiridomyrmecin

Dufour's glands:

Table 25. Hydrocarbons identified in the secretions of Dufour's gland of myrmicine ants

Species	Compound
Pogonomyrmex rugosus and *P. barbatus*	*n*-Dodecane
Pogonomyrmex rugostus and *P. barbatus*	3-Methylundecane
Pogonomyrmex rugosus	5-Methylundecane
Pogonomyrmex rugosus	6-Methylundecane
Novomessor cockerelli, Myrmica rubra, Pogonomyrmex rugosus, and *P. barbatus*	*n*-Tridecane
Pogonomyrmex rugosus and *P. barbatus*	3-Methyldodecane
Pogonomyrmex rugosus and *P. barbatus*	6-Methyldodecane
Novomessor cockerelli, Progonomyrmex rugosus, and *P. barbatus*	*n*-Tetradecane
Pogonomyrmex rugosus and *P. barbatus*	5-Methyltridecane
Pogonomyrmex rugosus and *P. barbatus*	3.5-Dimethyldodecane
Novomessor cockerelli, Myrmica rubra, Pogonomyrmex rugosus, and *P. barbatus*	*n*-Pentadecane
Pogonomyrmex rugosus and *P. barbatus*	6-Methyltetradecane
Pogonomyrmex rugosus and *P. barbatus*	3,4-Dimethyltridecane
Myrmica rubra	7-Pentadecene
Aphaenogaster longiceps	α-Farnesene
Novomessor cockerelli and *Myrmica rubra*	*n*-Hexadecane
Mirmica rubra	Hexadecene
Myrmica rubra	Homofarnesene
Novomessor cockerelli, Myrmica rubra, Solenopsis invicta, S. richteri, and *S. geminata*	*n*-Heptadecane
Myrmica rubra	*cis*-8-Heptadecene
Myrmica rubra	Heptadecadiene
Myrmica rubra	Bishomofarnesene
Novomessor cockerelli and *Myrmica rubra*	*n*-Octadecane
Myrmica rubra	9-Octadecene
Novomessor cockerelli and *Myrmica rubra*	*n*-Nonadecane
Myrmica rubra	9-Nonadecene

Table 26. Compounds identified in the mandibular gland secretion of *Iridomyrmex humilis*

Compound	Species
2,5-Dimethyl-3-propylpyrazine 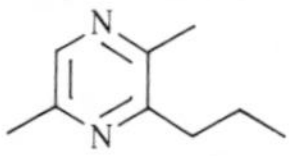	*Iridomyrmex humilis*
2,5-Dimethyl-3-isopentylpyrazine 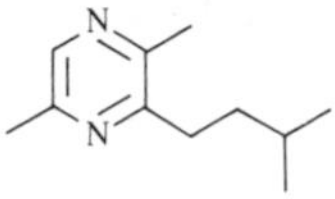	*Iridomyrmex humilis*
(E)-2,5-Dimethyl-3-styrylpyrazine 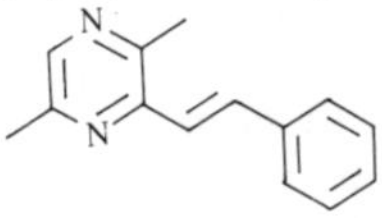	*Iridomyrmex humilis*

Anal Glands:

Table 27. Compounds identified in the anal gland secretions of dolichoderines

Compound	Species
2-Methylcyclopentanone 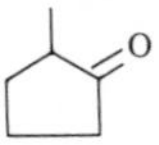	*Azteca* nr. *instubilis, A.* nr. *nigriventris,* and *A.* nr. *velox*
2-Heptanone 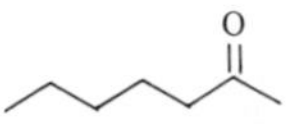	*Iridomyrmex pruinosus, Conomyrma pyramicus, Azteca chartifex, A. parzensis, A.* nr. *constrictor, A.* ssp. and *Monacis bispinosa*
4-Methyl-2-hexanone 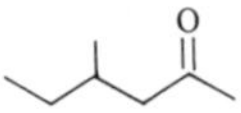	*Dolichoderus clarki*
6-Methyl-5-hepten-2-one 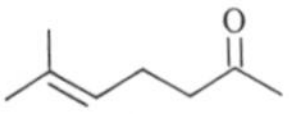	*Iridomyrmex detectus, I. conifer, I. rufoniger, I.* nr. *pruinosus, I. nitidiceps, Liometopum microcephalum, Conomyrma pyramicus, Tapinoma nigerrimum, Dolichoderus scabridus, Azteca chartifex, A. parzensis, Monacis bispinosa*

Compound	Species
2-Methyl-4-heptanone	*Tapinoma nigerrimum* and *T. sessile*
cis-1-Acetyl-2-methyl cyclopentane	*Azteca* nr. *velox* and *A.* nr. *nigriventris*
2-Acetyl-3-niethyl cyclopentene	*Azteca* nr. *nigriventris* and *A.* nr. *instabilis*
Acetic acid CH_3COOH	*Liometopum microcephalum*
n-Butyric acid	*Liometopum microcephalum*
Isovaleric acid	*Liometopum microcephalum* and *Iridomyrmex nitidiceps*
Benzaldehyde	*Azteca* sp.
Iridodial 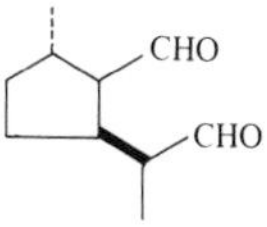	*Iridomyrmex detectus, I. conifer, I. pruinosus, Tapinoma nigerrimum, T. sessile, Dolichoderus scabridus, Conomyrma pyramicus, Azteca* nr. *instabilis, A.* nr. *velox, A. chartifex, A. parzensis, A.* ssp., and *Monacis bispinosa*

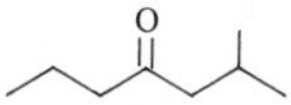

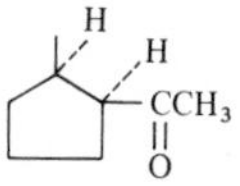

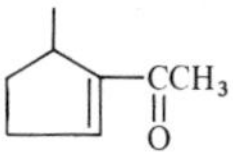

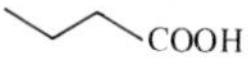

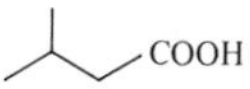

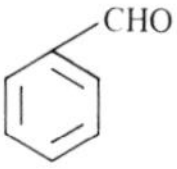

Table 27 (continued)

Compound	Species
Dolichodial 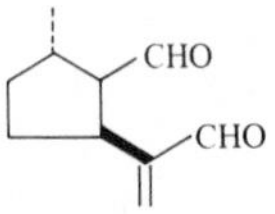	*Dolichoderus clarki, D. dentata, D. scabridus, Iridomyrmex detectus, I. rufoniger,* and *I. humilis*
CHO CHO	
Iridomyrmecin 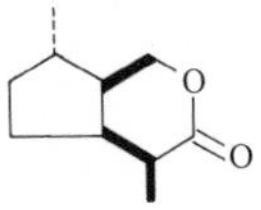	*Iridomyrinex humilis* and *I. pruinosus*
O O	
Isoiridomyrmecin 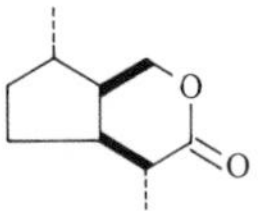	*Iridomyrmex nitidus, Dolichoderus scabridus,* and *Tapinoma sessile*
O O	
Isodihydronepetalactone 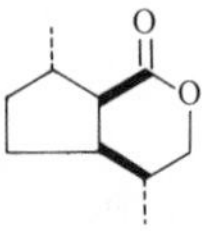	*Iridomyrmex nitidus*
O O	
2-Heptanol 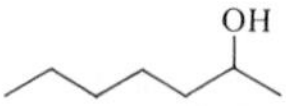	*Azteca* sp.
OH	
6-Methyl-5-hepten-2-ol 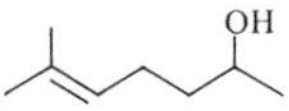	*Irodomyrmex* nr. *pruinosus*
OH	
2-Pentanone 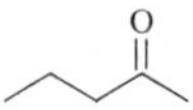	*Azteca* sp. and *Monacis bispinosa*
O	

Mandibular Glands

Ponerinae:

Table 28. Compounds identified in the mandibular gland secretions of ponerine species

Species	Compound
Paltothyreus tarsatus	Dimethyldisulfide
Paltothyreus tarsatus	Dimethyltrisulfide
Odontomachus hastatus	2,5-Dimethyl-3-isopentylpyrazine
Odontomachus brunneus	2,6-Dimethyl-3-*n*-pentyl-, *n*-butyl-, *n*-propyl-, and ethylpyrazine
Neoponera villosa	4-Methyl-3-heptanone
Neoponera villosa	4-Methyl-3-heptanol
Gnamptogenys pleurodon	Methyl 6-methylsalicylate

An extremely interesting defense mechanism is used by *Tapinoma nigerimum.* The defense secretion consists of the ketones methylheptenone and propyl-isobutyl-ketone and the dialdehyde Iridodial. The mixture is sprayed on the aggressor. Iridodial polymerizes on the enemy and immobilizes it. Additional volatilization of the toxic ketones is thus prevented.

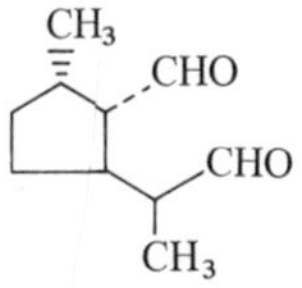

CH_3 CHO CHO CH_3

Iridodial

Closely related to Iridodial is the isomer Dolichodial from *Dolichoderus acanthoclinea:*

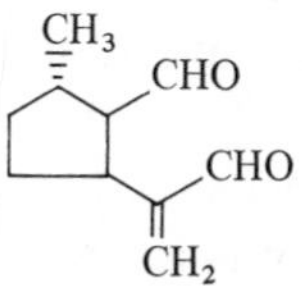

CH_3 CHO CHO CH_2

Dolichodial

References

Maschwitz, U. W., Kloft, W.: In "Venomous Animals and Their Venoms", Vol. III, (W. Bücherl and E. E. Buckley, Ed.) New York: Academic Press 1971
Buffkin, D., Russell, F. E.: In: "Tier- und Pflanzengifte" (E. Kaiser, Ed.) Munich: Goldmann 1973
Brand, J. M., Blum, M. S., Fales, H. M., Maclounell, J. G.: Toxicon *10*, 259 (1972)
Maschwitz, U., Koob, K., Schildknecht, H.: J. Insect Physiol. *16*, 387 (1970)
Stumper, R.: Naturwissenschaften *47*, 460 (1960)
Pavan, M.: Chim. e Industr. *37*, 625 (1955)
Fusco, R., Trave, R., Vercellone, A.: Chim. e Industr. *37*, 251, 958 (1955)
Doljejs, L., Mironow, A.: Tetrahedron Letters *11*, 18 (1960)
Bates, R., Eisenbraun, E., McElvain, S.: J. Amer. Chem. Soc. *80*, 3420 (1958)
Cavill, G., Ford, D.: Austr. J. Chem. *13*, 296 (1960)
Korte, F., Falbe, J., Zschocke, A.: Tetrahedron *6*, 201 (1959)
Clark, K., Fray, G., Jaeger, R., Robinson, R.: Tetrahedron *6*, 217 (1958)
Clark, K., Fray, G., Jaeger, R., Robinson, R.: Angew. Chem. *70*, 704 (1958)
Trave, R., Pavan, M.: Angew. Chem. *70*, 115 (1958)
Cavill, C., Ford, D.: Chem. u. Industrie 351 (1953)
Cavill, C., Hinterberger, H.: Austr. J. Chem. *14*, 143 (1961)
Meinwald, J., Chadhe, M., Hurst, J., Eisner, T.: Tetrahedron Letters *1*, 29 (1962)
Bradshaw, J. W. S., Baker, R., Howse, P. E.: Nature *285*, 230 (1962)
Mac Connell, J. G., Blum, M. S., Buren, W. F., Williams, R. N., Fales, H. M.: Toxicon *14*, 69 (1976)
Blum, M. S., Hermann, H. R.: In "Arthropod Venoms" (Ed. S. Bettini), 1978

3.3.6 Poisonous Insects

In addition to the actively venomous insects described so far there are a number of secondarily venomous or poisonous species. They are special in that they do not produce their venoms themselves, but obtain them from plants upon which they feed. This protection is not unlimited. A review is given in Table 29:

These animals live as caterpillars on toxic plants and take up these toxins together with their food. The toxins are not metabolized further, but stored in the organism. Larvae and adults stay venomous and thus are protected against potential enemies. The use of those substances is so great that even the eggs of these insects contain considerable amounts of venoms and thus are also protected.

Table 29. Poisonous insects by plant venoms

Order or species	Food plant	Toxin obtained
ORTHOPTERA		
Pyrgomorphidae		
6 species	Asclepiadaceae	Cardenolides
HEMIPTERA		
Heteropteroidea		
Aphididae		
Aphis nerii	Nerium oleander	Cardenolides
	Asclepias curassavica	Calotropin
Lygaeidea		
35 species	Nerium oleander	Cardenolides
Sternorrhyncha		
Diaspididae		
Aspidiotus nerii	Oleander	Cardenolides
NEUROPTERA		
Chrysopidae		Cardenolides
COLEOPTERA		
Cerambycidae		
Tetraopes spp.	Asclepiadaceae	Cardenolides
Chrysomelidae		
Chrysolina brunsvicensis	Hypericum spp.	Hypericine

References

Eisner, T.: "Chemical Ecology", p. 157 (E. Sondheimer and J. B. Simeone, Eds.) New York: Academic Press 1970

Blum, M. S.: Pesticide Chemistry *3*, 147, 163 (1971)

Rothschild, M.: In: "Insects/Plant Relationships" (H. F. van Emden, Ed.) Symposia of the Royal Entomological Society, No. 6 Oxford: Blackwell Scientific Publications, 1972

Rothschild, M., Euw, J. v., Reichstein, T.: Insect. Biochem. *2*, 344 (1972)

Rothschild, M., Euw, J. v., Reichstein, T.: Proc. Roy. Soc (London) B, *183*, 227 (1973)

4 Echinodermata (Echinoderms)

Echinoderms are marine invertebrates possessing a characteristic fivefold symmetry. Their skeleton consists of calcium carbonate lamina of variable size. The skin is firm and cast with spines, thorns or pedicellariae (small prehensile pincers). The phylum *Echinodermata* comprises about 5,300 species. It is divided into two subphyla:

1. *Pelmatozoa*
2. *Eleutherozoa*

Pelmatozoa have scarcely been investigated and venomous species have not yet been found. The *Eleutherozoa* are subdivided into four classes:

a) *Holothurioidea* (sea cucumbers)
b) *Crinoidea* (sea urchins)
c) *Asteroidea* (starfishes)
d) *Ophiuroidea* (brittle stars)

The echinoderms are distributed in all seas from the tropics to the arctic zones.

4.1 Holothurioidea (Sea Cucumbers)

As the name implies, the sea cucumber possesses an enlongate body. Feeding on the bottom of the oceans, they live from the organic material. Their venom therefore is not used for predation but rather in self defense. Sea cucumbers are widely distributed; they populate the coastal areas of practically all tropical and subtropical seas.

Poisoning and Treatment

Like the starfishes, sea cucumbers possess small skin glands excreting a venom. A more important source of envenomation, however, are the Cuvier's tubules, which may be found in many but not all species (Fig. 11). When the animal is endangered, they are ejected from the body cavity through the anus and upon contacting water swell and enlongate to sticky threads. Envenomation therefore occurs primarily after consumption of holothurians, if during preparation of "Trepang" or "Bêche-de-mer" the venoms have not been removed by thorough washing and watering of the animals.

The LD_{50} is 7.5 mg/kg (mouse, i. v.); a dilution of 1 : 100000 of the crude toxin was found to kill fishes within a few minutes. The consumption of holothurian venoms causes, in mild cases, digestive

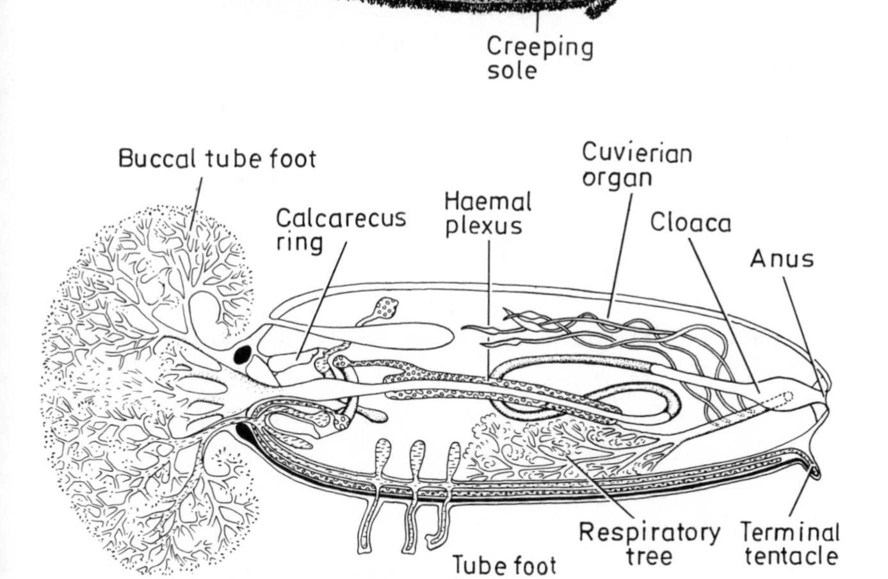

Fig. 11. Anatomy of a typical sea cucumber (from Halstead)

problems; in serious cases paralysis and death have been observed. There is no reliable method for treatment; all measures have to be taken symptomatically.

Envenomation by injury to human skin or infiltration of the venom from the skin gland will cause burning pains and inflammation, which may be treated symptomatically; they usually need no treatment, however.

Chemistry

The chemistry of holothurinotoxins has been the object of intense investigation in recent years; the chemical structure of the individual compounds has been elucidated in detail. There is no difference between the toxins from Cuvier's glands and those from the skin. They are glycosides, the aglycones of which possess a modified lanosterol skeleton. The species investigated are listed in Table 30.

Table 30. Investigated species of sea cucumbers

Species	Author
Actinopyga agassizi	Chanley 1966, 1969a, b
Halodeima grisea	Tursch, 1967
Holothuria tubulosa	Habermehl, 1968
Holothuria forskali	Habermehl, 1968
Holuthuria polii	Habermehl, 1970
Bohadschia koellikeri	Roller, 1969; Tursch, 1970
Stichopus japonicus	Elyakov, 1969

Some of the typical aglyca are represented in the following formula:

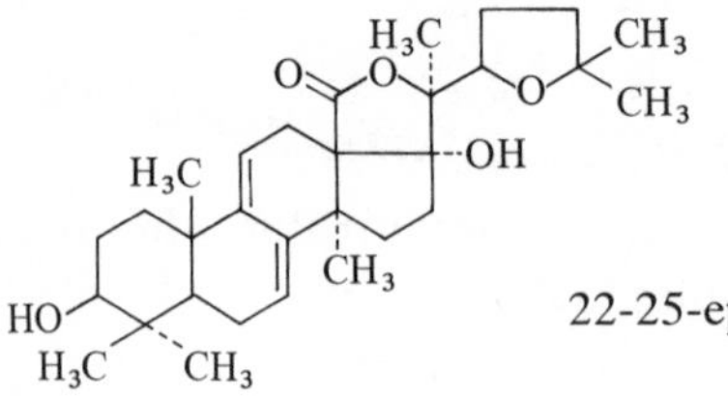

22-25-epoxy-holothurinogenin

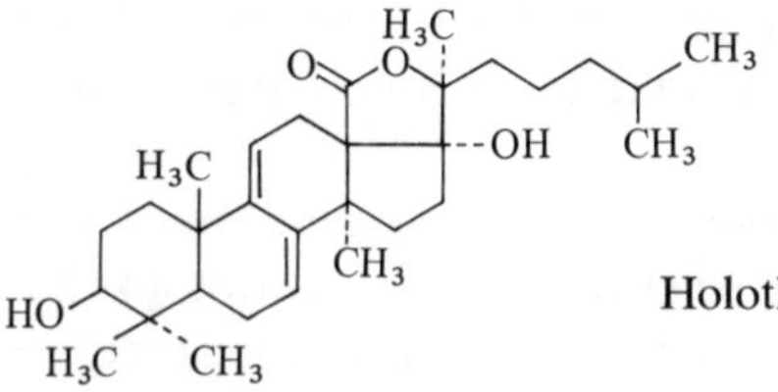

Holothurinogenin

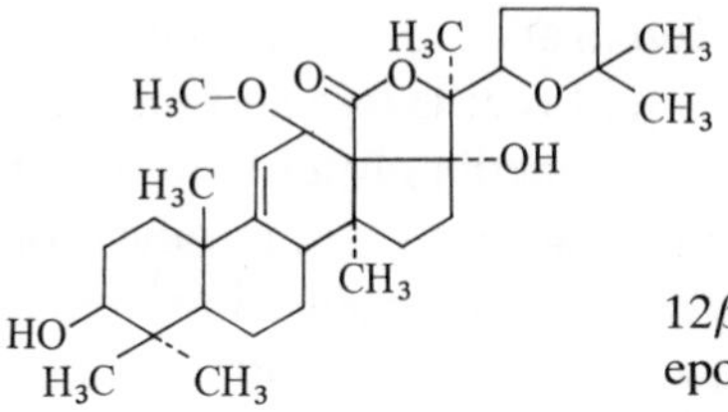

12β-methoxy-7.8-dihydro-22.25-epoxy-holothurinogenin

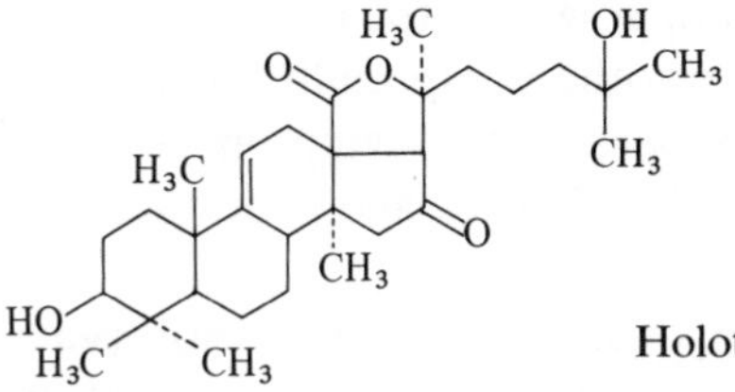

Holotoxin

4.2 Echinoidea (Sea Urchins)

Sea urchins are well known from tropical or subtropical seas. They are responsible for most incidents involving *Echinodermata.* Nine families with many venomous species are found in the coastal areas of Europe, Asia, Australia, the Atlantic and Pacific Islands and of Central America.

Poisoning and Treatment

Poisoning can occur in two ways. Most of the sea urchins are venomous during their mating season, since the toxins are produced in their genital glands. Such animals should not be eaten unless one has been assured that they are safe. The usual symptoms are nausea, vomiting and diarrhea; strong headache and allergic reactions have also been described. Although some rare fatal cases have been reported, the envenomations usually take a benign course and end without complications.

A second route of envenomation is via the spines, which can produce painful wounds. In most species the resulting symptoms are merely of mechanical origin, from the broken parts of the spines in the skin. In the families *Echinothuridae* and *Diadematidae,* however, true envenomation can be observed in addition. The spines are tubular and filled with a violet-color liquid that is injected into the skin after the sting. Extreme pains occur immediately and the site of the sting becomes reddish and swollen; in severe cases paralysis of the motor nerves of the affected limb and irregular pulse have been observed. Pains diminish within a few hours, but the other symptoms may last for 3–4 days.

Envenomations can occur in a similar way by the pedicellariae of the families mentioned above and of others, especially *Toxopneustidae.* In addition to the local reactions, general symptoms such as paresthesia, numbness, respiratory disturbances and even fatalities have been reported from *T. pileolus.*

Treatment must focus on the removal of all remainders of spines and pedicellariae from the skin, surgically if necessary. All other effects should be treated symptomatically, as there is no specific therapy.

Chemistry

There has been little investigation of the chemistry of sea urchin toxin. The toxin of *Tripneustes gratilla* (family *Toxopneustidae*) is a pH- stable, non-dialyzable, thermolabile protein. It is completely soluble in distilled water and can be precipitated with ammonium sulfate. The crude toxin can be separated into six components on an acrylamide gel. The structure of the compounds has not yet been determined.

4.3 Asteroidea (Starfishes)

Poisoning from starfishes is long known. The older literature is full of speculations and only partially correct facts. Starfishes produce in their skin glands a slimy secretion that is delivered into the water. Using these toxins they are able to paralyze and hunt shellfish, snails and small shrimps.

Poisoning and Treatment

The crude venom irritates the mucous membranes of higher animals; swelling and inflammation may be observed. An indeterminate cardiac activity has also been reported. Care should be taken in handling starfish to avoid injury to the skin from the spines. If the toxin penetrates the skin a painful dermatitis will result. The oral route of poisoning is dangerous. After only a few minutes paralysis occurs followed by convulsions and then death after a few hours. Treatment is symptomatic.

The venom not only serves active predation but also possessess antibiotic activity, especially against sea water resistant microorganisms. Since the roe is also toxic, and this toxin is obviously identical to the skin toxin, there seems to be a further protective function.

Chemistry

Thorough investigations have been carried out since 1960. The species *Acanthaster planci, Marthasterias glacialis, Asterias amurensis, Asterias rubens* and *Solaster papposus* have been studied by many groups all over the world. The results show the purified toxins to be steroid glycosides. The following are typical aglyca:

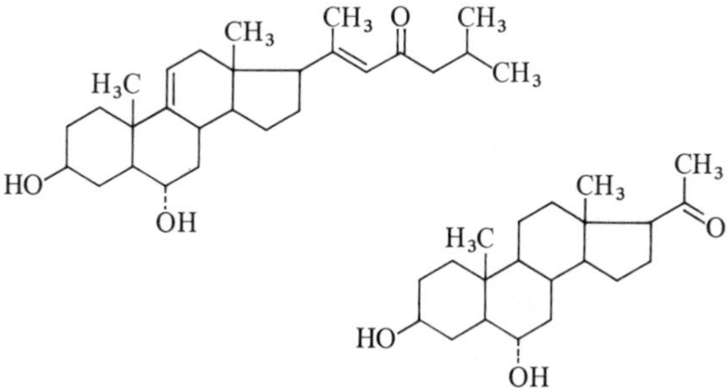

CH_3 CH_3 O CH_3 H_3C CH_3 HO OH

CH_3 CH_3 H_3C O HO OH

The hydroxy-group at C-3 occasionally is esterified with sulfuric acid:

CH_3 CH_3 H_3C O $^{\ominus}O_3S$ O OH

The sugar moiety is attached to the hydroxy group at C-6 of the steroid.

The starfish, *Acanthaster planci,* widely distributed in the Pacific, has become well known during recent years as a menace for coral reefs. Attempts to cope with this have been of little success so far.

4.4 Ophiuroidea (Brittle Stars, Serpent Stars)

Relatively little is known about these animals. The class *Ophiuroidea* consists of two families, *Gorgonocephalidae* and *Ophiocomidae.* They very much resemble the starfishes but possess five long, thin and fragile arms, which, in constrast to the starfishes, are separated distinctly from each other. They feed upon shellfish, snails and small shrimps. Envenomation in humans has not been reported, although it is known from pharmacological tests that the venom from the spines is capable of paralyzing and ultimately killing small mammals.

Chemical investigations have shown that these compounds are also steroid glycosides.

References

Summary

Halstead, B. W.: Poisonous and Venomous Marine Animals, Vol. 1, U. S. Govt. Printing Office, Washington, D. C. 1965

Starfishes

Hashimoto, Y., Yasumoto, T.: Bull. Jap. Soc. Sci., Fish *26,* 1132 (1960)

Rio, G. J., Ruggieri, G. D., Stempien, Jr. M. F., Nigrelli, R. F.: Am. Zool. *3,* 554 (1963)

Yasumoto, T., Tanaka, M., Hashimoto, Y.: Bull. Jap. Soc. Sci., Fish *32*, 673 (1966)
Mackie, A. M., Turner, A. B.: Biochem. J. *117*, 543 (1970)
Goad, L. J., Rubinstein, I., Smith, A. G.: Proc. Roy, Soc. (London), Ser. B *180*, 223 (1972)
Sheik, Y. M., Tursch, B., Djerassi, C.: J. Amer. Chem. Soc. *94*, 3278 (1972)
Turner, A. B., Smith, D. S., Mackie, A. M.: Nature (London) 233, 209 (1971)
Ikegami, S., Kamiya, Y., Tamura, S.: Tetrahedron Letters (London) 1601 (1972)
Gupta, K. C., Scheuer, P. J.: Tetrahedron (London) *24*, 5831 (1968)
Kobayashi, M., Tsuru, R., Todo, K., Mitsuhashi, H.: Tetrahedron Letters (London) 2935 (1972)
Smith, A., Goad, L. J.: Biochem. J. *123*, 671 (1971)
Smith, A. G., Rubinstein, I., Goad, L. J.: Biochem. J. *135*, 443 (1973)
Preusser, H. J., Habermehl, G., Sablofski, M., Schmall-Haury, D.: Toxicon *13*, 285 (1975)

Sea cucumbers

Chanley, J. D., Mezetti, T., Sobotka, H.: Tetrahedron *22*, 1857 (1966)
Chanley, J. D., Rossi, C.: Tetrahedron *25*, 1897 (1969a)
Chanley, J. D., Rossi, C.: Tetrahedron *25*, 1911 (1969b)
Elyakov, G. B., Kuznetsova, T. A., Dzizenko, A. K., Elkin, Y. N.: Tetrahedron Letters 1151 (1969)
Friess, S. L., Standaert, F. G., Withcomb, E. R., Nigrelli, R. F., Chanley, J. D., Sobotka, H.: Pharmac. exp. Therap. *126*, 323 (1959)
Friess, S. L., Standaert, F. G., Withcomb, E. R., Nigrelli, R. F., Chanley, J. D., Sobotka, H.: Ann. N. Y. Acad. Sci *90*, 893 (1960)
Friess, S. L.: A. I. B. S. Bull. *13*, 41 (1963)
Friess, S. L., Durant, R. C., Chanley, J. D., Mezetti, R.: Biochem. Pharmac. *14*, 1237 (1965)
Friess, S. L., Durant, R. C., Chanley, J. D., Fash, F. J.: Biochem. Pharmac. *16*, 1617 (1967)
Friess, S. L., Durant, R. C.: Toxicon. appl. Pharmac. *7*, 373 (1965)
Friess, S. L., Durant, R. C., Chanley, J. D.: Toxicon 6, 81 (1968)
Habermehl, G., Volkwein, G.: Naturwissenschaften *55*, 83 (1968)
Habermehl, G., Volkwein, G.: Liebigs Ann. Chem. *731*, 53 (1970)
Nigrelli, R. F.: Zoologica *37*, 89 (1952)
Nigrelli, R. F., Jakowski, S.: Ann. N. Y. Acad. Sci. *90*, 884 (1960)
Roller, P., Djerassi, C., Cloetens, R., Tursch, B.: J. Amer. Chem. Soc. *91*, 4918 (1969)
Tursch, B., Souza Guimaraes, I. S. de, Gilbert, B., Aplin, R. T., Duffield, A. M., Djerassi, C.: Tetrahedron *23*, 761 (1967)
Tursch, B., Cloetens, R., Djerassi, C.: Tetrahedron Letters 467 (1970)
Yamanouchi, T.: Publ. Seto mar. Biol. Lab. *4*, 183 (1955)
Kitagawa, J., Sugawara, T., Yosioka, J.: Tetrahedron Letters 963 (1975)
Tan, W. L., Djerassi, C., Fayos, J., Clardy, J.: J. Org. Chem *40*, 466 (1975)
Kelecom, A., Daloze, D., Tursch, B.: Tetrahedron *32*, 2353 (1976)

5 Pisces (Fishes)

5.1 Poisonous Fishes

More than 500 species have been reported to cause poisoning among humans when eaten. Such poisonings are well known since antiquity. A hieroglyphic epigraph on the tomb of the Pharaoh Ti (2500 B. C.) illustrated the toxicity of the buffer fish *(Tetraodon spp.)*.

The distribution of poisonous fishes favors the tropical seas. The toxicity of individual species is variable. Some fishes are poisonous throughout the year whereas others are poisonous at certain times only. Some species are entirely poisonous, in others the toxicity is restricted to certain organs. These fishes are referred to as *ichthyosarcotoxic.* A further differentiation is made acording to the symptoms caused by the different toxins.

5.1.1 Ciguatera-toxic Fishes

The toxic manifestations of ciguatera envenomations are gastrointestinal and neurotoxic symptoms. They may appear after ingestion of barracuda, sea perch, doctor fishes, snapper and parrot fish. Most of these fishes are usually edible, they are even considered economically valuable; hence, envenomations occur again and again. About 300 species of the families *Acanthuridae, Aluteridae, Balistidae, Carangidae, Chaetodontidae, Labridae, Lethrinidae, Lutjanidae, Muraenidae, Scaridae, Serranidae* and *Sphyraenidae* are known to be responsible for ciguatera poisoning.

It is evident from many studies that ciguatera toxin is not produced by the fishes themselves. It is taken up via food and then stored in distinct parts of the body, primarily in liver, testes and intestines; the meat is generally less toxic. The toxin is produced by algae. Ciguatera poisoning is widely distributed in the Pacific Islands, but also on the East and West Coasts of the USA.

Poisoning and Treatment

The initial symptoms usually occur within the first four hours after consumption of the fish: painful tingling about the mouth, nose and throat, occasionally also in fingers and toes, nausea and vomiting, abdominal pain and diarrhea. Later on general weakness, chills, pruritus and rapid fatigue, fever, bradycardia and hypotension, insomnia, headache and backache, respiratory distress and myalgia "particularily in the legs" can be observed. Convulsions precede death.

Ciguatera toxin does not produce immunity. A second envenomation within six months produces a distinctly more severe course than the first one.

Treatment in most of the minor cases can be symptomatic. Among the natives of New Caledonia a tea from the leaves of a plant *Duboisia myoporoides* has been used successfully – even in severe cases. Recent investigations on the chemistry of this plant showed that the active principle is a mixture of tropa and pyridine alkaloids, that is nicotine, nor-nicotine, atropine and scopolamine.

Atropine also is the drug of choice for controlling the symptoms during the early stage of the poisoning. Steroids are ineffective. Calcium gluconate, vitamin B complex and supplemental vitamin C coupled with a high protein diet gave most satisfactory results. An analgesic cream can be used to control pruritus. Diazepam can be advised against insomnia. Hydrocortisone, neostigmine and chlorpromazine, noted elsewhere and in earlier studies of this poisoning, have shown to be of no value.

Chemistry

Ciguatera toxin shows the properties of a cholinesterase inhibitor, but in spite of intense investigations the structure is not yet clear. It is definitely not a protein or a peptide. The substance isolated in 0.0006% yield from a pacific snapper, *Lutjanida bohar,* by a long process of extractions and refinement is soluble in the usual organic solvents (e. g. ether, petroleum ether, acetone, chloroform, methanol and 90% ethanol). It is insoluble in water. The toxin is a neutral substance with a LD_{50} of 0.08 mg/kg (mouse, i. p.).

5.1.2 Tetrodo-toxic Fishes

This group comprises all fishes containing tetrodotoxin. More than 50 species of the families *Diodontidae* (urchin-fishes), *Tetraodontidae* (Fugu, puffer fishes, bowl fishes) and *Molidae* (sun fishes) are known to be poisonous.

Tetrodotoxin poisoning has been known for a long time, and it is considered one of the important envenomations in view of the number as well as the severity of the cases and especially since these fishes are considered a delicacy in Japan and in South East Asia. In Japan they may only be prepared by licensed cooks in special restaurants. Nevertheless more than 150 fatalities are reported there each year, and considering East Asia and the Pacific area the number of deaths is much higher. The LD_{50} is 8 μg/kg (mouse, s. c.), and the mortality rate in humans is 60%. Thus tetrodotoxin is among the strongest non-protein toxins.

Tetrodotoxin is found in the ovaries and testes, liver and intestines; the concentration in meat and skin is lower. The toxicity of the fish depends on the reproductive cycle; it is highest shortly before spawning, ie, from May to July.

Poisoning and Treatment

Poisoning is characterized by the rapid appearance of the symptoms within 5 to 30 minutes after ingestion. These are general weakness, dizziness, nausea, faintness, pallor, paresthesia and prickling on lips, tongue and throat. Later on this prickling sensation also seizes the fingertips and toes. Contrary to widespread opinion, tetrodotoxin is not an emetic, and vomiting is, despite nausea, very rare; if it occurs at all it is within the first hour after poisoning. In an advanced stage sweating, respiratory distress and hypotension set in. In severe cases myalgia, pains in the chest and cyanosis are also observed. In the last stages paralysis and occasionally convulsions appear. Death is due to respiratory paralysis within 6–24 hours after poisoning.

Though the pharmacology of tetrodotoxin has been the subject of extensive, intense investigation, there is no specific antidote; poisoning can only be treated symptomatically. Artificial respiration and cardiac massage may be necessary.

Chemistry

Attempts to isolate and to elucidate the structure of tetrodotoxin were made starting at the end of the last century, particularly by Japanese scientists. The first investigations were by Tahara, in 1894; the final structure determination was done by Tsuda and Goto in 1963.

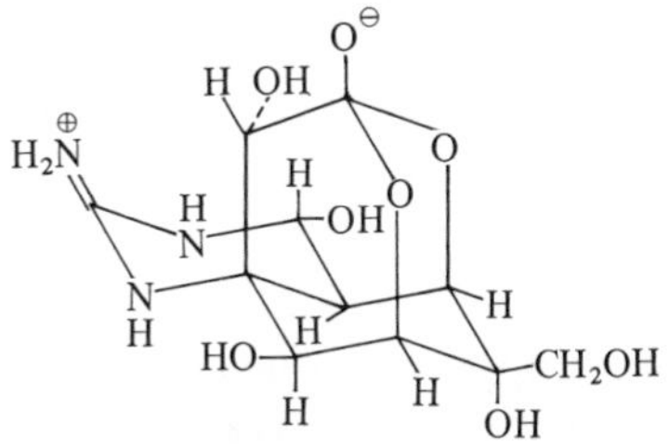

Tetrodotoxin

Tetrodotoxin not only acts on the nerve endings, but also directly on the heart. The actions are similar to those of some local anesthetics. It is procaine and cocaine in that it selectively blocks the sodium-ion transport through cell membranes without affecting the permeability for potassium-ions. In this respect it is the antagonist of Batrachotoxin (cf. p. 122).

5.1.3 Ichthyoo-toxic Fishes

There are a number of freshwater fishes as well as marine species that contain a toxin restricted to the gonads and roe. The fishes themselves are edible without danger.

Poisoning

Poisoning usually produces no serious consequences. Recovery takes place within a few days without treatment. Symptoms are nausea, vomiting and diarrhea; in the rare severe cases respiratory distress, pains in the chest and convulsions have been observed.

Chemistry

Unknown

5.1.4 Ichthyohemo-toxic Fishes

The blood serum of many species of the eels *Anguilla* and *Muraena* contains a protein having a hemolytic action on erythrocytes of warm-blooded animals. If raw blood of these eels is consumed, nausea, vomiting, diarrhea, eruption and general weakness appear. In severe cases respiratory distress and paralysis have been observed; several fatalities have even been reported.

Treatment is symptomatic.

5.1.5 Other Poisonings

The fish poisonings described so far are not the only ones that have been reported. Other cases have been attributed to certain species, usually depending on season or region, and they are generally associated with headache, digestive troubles and muscular weakness. The symptoms usually occur soon after ingestion and disappear spontaneously within 24 hours.

It must be mentioned, however, that after consumption of shark liver severe envenomations, even fatalities, have been observed. Poisoning is characterized by nausea, vomiting, headache and abdominal pain and diarrhea. The pulse is weak and rapid, and visual disturbances have occasionally been described. Respiratory distress and coma precede death. The time for complete recovery is 5 to 20 days. There is no certain explanation for this syndrome, but it is probably a vitamin A-hypervitaminosis.

5.2 Actively Venomous Fishes

About 250 species of fishes are actively venomous; that is, they possess a venom apparatus consisting of glands and stings used for defense. The most important species will be described here.

5.2.1 Dasyatidae (Stingrays)

There are many different sizes of stingrays; they range 10 cm in diameter (body) up to one meter or wider. The length including the tail in the largest species may reach 4 meters. Stingrays live in the shallow waters (up to 30 m depth) close to the shore, but there are also some freshwater stingrays found in South American rivers. They frequently hide in the sand, and accidents occur when bathers step on them and are injured by the sting, which is located on the dorsal side of the last third of the tail.

The flattened sting has sawtooth-like barbs on both sides and may be 4 cm to 30 cm in length (df. fig. 12). Wounds are large and painful because of this structure. The toxin enters easily, and secondary infections are common. 25% of the patients need surgical

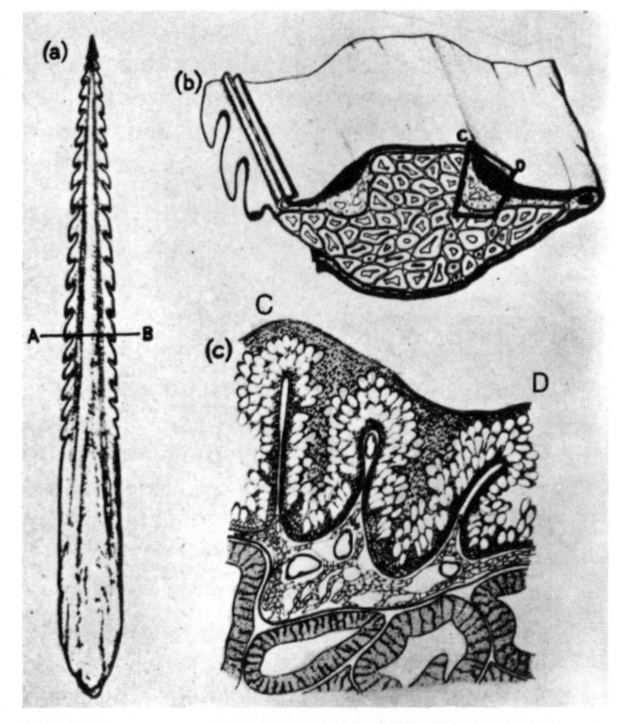

Fig. 12. Venom apparatus of a sting ray *(Urolphus halleri)*. **a** sting with barbed hooks, **b** profile at the site *A–B*, **c** square beneath *C–D* from b enlarged; the venom producing cells are visible in the epithelium (from F. E. Russell)

treatment of the wound. About 1% of the cases end in death. Exact statistics on the frequency of incidents are not available; along the coast of the USA, however, about 750 cases per year are registered. The figures for the Pacific, Indo-Pacific, Mediterranean Sea and South Atlantic are probably higher.

Poisoning

The integumentary sheath of the sting is destroyed when the sting penetrates the limb, and the toxins can enter the wound. Upon withdrawal of the sting, some of the integumentary sheath usually remains. In any case the injury demands careful and intensive treatment. Wounds may be as long as 15 cm. The pains are much more intense than would be expected from a simple wound; they increase in intensity for 30 to 90 minutes and last for 6 to 48 hours. The area immediately surrounding the wound is discolored, and in untreated cases necrosis of this region develops. The general condition of the victim is usually bad: faintness, nausea and feelings of fear, sometimes collapse can occur and are attributable to peripheral vasodilatation. Vomiting and intestinal disturbances are common. Cramps and paralysis have very rarely been observed. They are not directly caused by the toxin, but result from the excruciating pains.

Treatment

Successful treatment requires prompt action. Analgesics must be given and all necessary steps taken to prevent secondary infections, which can be controlled with antibiotics. Protection against tetanus is also important.

Treatment of the wound should start with thorough washing to remove at least part of the toxin. If nothing else is available even sea water can be used. The afflicted limb should be bathed in water as hot as possible (without scalding) for 30 to 90 minutes. The toxin is heat labile and thus can be easily inactivated. Then the wound must be cleaned (residues of the integumentary sheath have to be removed), desinfected and sutured.

But the reader is warned never to treat the wound with potassium permanganate, liquid ammonia, formaldehyde, soft soap or ice water. All of these substances will cause more harm than not treating the wound at all.

Chemistry

The aqueous solution of the crude toxin is clear and colorless; the pH value is around 7. A faint smell of ammonia may be noted. Toxicity is lost at room temperature within 18 hours due to the thermolability. Serotonin, 5-nucleotidase and phosphodiesterase, and three protein fractions of high specific toxicity can be isolated in addition to free amino acids. The molecular weight is around 100,000. Protease or phospholipase have not been detected.

5.2.2 Trachinidae (Weaver Fishes)

The weaver fishes (family: *Trachinidae)* are distributed along the coasts of the East Atlantic and the Mediterranean Sea. The name "weaver" probably is a corruption of the word "viper". The most important species are the smaller weaver (*Trachinus vipera, Viperqueise)* and the greater weaver (*T. draco).* T. vipera is abundant in the Mediterranean Sea, along the South coast of England and in the Southern North Sea. Because of its small size (15 cm) it has no commercial value. *T. draco,* however, native to the North Sea as well as the Mediterranean Sea, reaches a size up to 45 cm, and is highly regarded as a food fish.

The sting apparatus consists of two opercular spines and five to eight dorsal spines. At the base of the spines a spongy, glandular tissue is found that produces the venom. The spines are covered with an integumentary sheath that ruptures during the sting enabling entry of the venom into the wound.

Poisoning and Treatment

Poisoning can occur in different ways. In 80% of all cases fishermen are stung while removing fish from their nets or unhooking them. *T. vipera,* moreover, have the habit of digging themselves into the sandy bottom, so that bathers may step on them.

Both *T. vipera* and *T. draco* use their spines primarily for defense; but there are many reports from skin divers that *T. draco* especially if threatened will use its spines for attack. Since these attacks are sometimes directed against bigger fishes, and since the weaver does not follow its victim for any distance, it seems likely that this behavior serves territorial defense. The sting is extremely painful, with pains intensifying for about 24 hours, spreading out over the whole limb. Bleeding from the wound is not uncommon. Symptoms and treatment are as described for stingrays. Hot water treatment is strongly recommended. Improper treatment with soft soap or potassium permanganate, practised on trawlers for a long time, leads to the loss of the limb involved.

5.2.3 Scorpaenidae (Scorpionfishes, Rockfishes)

About 80 species of the family Scorpaenidae are involved in the accidental poisoning of humans. They are distributed in all tropical and subtropical seas; even in the arctis regions, individual species can be found. Some species are eye-catchingly colored whereas others blend into their environment; still others dig themselves into the bottom. They can be divided into three groups according to their venom apparatus.

Table 31. Classification of scorpionfishes (according to Halstead)

Structure	*Pterois*	*Scorpaena*	*Synanceja*
Fin spines	Elongated and Slender	Shorter and heavier than in Pterois	Short and stout
Integumentary sheath	Thin	Thick	Very thick
Venom glands	Small but well developed	Large and highly developed	Large and highly developed
Venom ducts	None	None	Well developed

5.2.3.1 Pterois, Dendrochirus (Zebrafishes, Butterfly Cods)

They belong to the most spectacularly colored fishes living in the shallow waters around coral reefs. These species are relatively infrequently involved in accidents; they are, however, capable of producing fatalities in humans. The venom apparatus consists of 13 dorsal spines, 2 pelvic spines and 3 anal spines covered with integumentary sheaths in which the venom glands are embedded.

Poisoning and Treatment

Poisoning occurs as a result of mechanical pressure on the spines, for example upon carelessly grabbing a fish. The sting immediately causes intense burning pains that spread rapidly and sometimes become so excruciating that the victim may become unconscious. Dizziness, fainting, decrease of the heart rate to less than 50 beats per minute and shock may quickly ensue. Hypothermia and respiratory distress may last for 12 hours. The sting site is swollen and discolored, and may remain so for weeks. Usually the area around the wound becomes necrotic.

Hot water treatment has proved to be useful, although envenomation from *Pterois* usually is more severe and considerably more dangerous than poisonings from weaverfishes. Symptomatic treatment of the general symptoms has to be carefully observed.

Chemistry

The toxin is a thermolabile protein. At −20 °C the activity is reduced to 50% within one year. The LD_{50} is 1.1 mg/kg (mouse, i. v.).

5.2.3.2 Scorpaena (Scorpionfishes, Sculpins, Waspfishes, Blobs)

This subfamily is widely distributed and comprises the genera *Apistus, Centropogon, Gymnapistes, Hypodytes, Nothestes, Scorpaena, Scorpaenodes, Scorpaenopsis, Sebastapistes, Sebastodes, Sebastolobus* and *Snydcrina.* Their sting apparatus differs from that of the zebrafishes in that it has 12 instead of 13 spines, which are also shorter and thicker.

Poisoning and Treatment

Sculpin poisoning is very common; about 300 cases per year are estimated in the USA alone. 80% of the stings are inflicted upon fishermen and bait handlers.

The local and systemic symptoms are very similar to those

described for the zebrafishes. Nausea, vomiting, weakness, conjunctivitis, perspiration, headache and diarrhea have been reported. The best treatment is hot water. In severe cases the victim should be taken to the hospital for treatment of the general condition.

5.2.3.3 Synanceja (Stonefishes, Devilfishes, Sea Toad, Lumpfish)

This subfamily comprises the genera *Synanceja, Choridactylus, Erosa, Inimicus, Leptosynanceja* and *Minosus.* These fishes are probably the most dangerous and venemous piscines of all. The venom apparatus differs from those of the fishes mentioned so far. The spines are shorter and thicker and contain the oval venom gland in the upper third. The stonefishes are sluggish and frequently lie in the bottom in the shallow waters or even dig themselves in. Since they possess bizarre forms as well as the coloration of the sea floor, good and trained eyes are necessary to detect them.

Poisoning and Treatment

Stonefish stings produce a poisoning, the course of which is similar to those of other *Scorpaenidae,* but more severe. Fatalities are not at all rare, especially with *Synanceja horrida* and *S. trachinus;* two out of three cases are fatal within 6–8 hours. Treatment therefore has to start as soon as possible.

Locally, there are excruciating pains immediately after the sting. Though the wounds, depending on the size of the spine, are relatively large, there is little or no bleeding. The sting site acquires a blue discoloration within a few minutes. The originally local pains spread out over the whole limb and hardly respond to analgesics; shivers and profuse sweating set in. The afflicted limb becomes swollen. The swelling may last for weeks, and the wound does not heal. A greenish secretion exudes from it. Necrosis and formation of a tumor in the open wound may occur.

Among systemic effects, pains of the kidney and heart may last for weeks, too.

Treatment utilizing hot water has been shown to be successful. Natives in the Pacific region treat such wounds by pouring warm coconut oil on it. Both procedures have the same purpose: To destroy the heat-labile venom. Injection of emetine hydrochloride (0.01 g) directly into the wound is of value, if done within the first

30 minutes. Application of the very effective antivenin is strongly recommended, if the patient is not sensitive to horse serum. In any case the wound has to be treated surgically to clean and remove the residues of the integumentary sheath. Secondary infections can be controlled by antibiotics; tetanus vaccination is indicated.

Chemistry

The crude venom is a clear colorless fluid containing 13% protein. LD_{50} is 0.2 mg/kg (mouse, i. v.). The toxicity of the toxin itself can be seen from Table 32.

Table 32

Animal	i. v.	s. c.	i. p.
Mouse	0.01	0.04	0.02
Rabbit	0.01	–	–

figures given in mg/kg body weight

The toxin consists of a mixture of 10 proteins, only one of which, however, is lethal. The toxicity is lost within 48 hours if the venom is stored in solution at room temperature. Lyophilized venom is stable for several months. The toxin is inactivated by acids, ammonia, alcohol and heat. There are no investigations about the structure of the protein.

References

Halstead, B. W.: Poisonous and Venomous Marine Animals, US. Govt. Printing Office, Washington, D. C., Vol. 2, 1967; Vol. 3, 1970

Halstead, B. W.: In "Venomous Animals and Their Venoms" (W. Bücherl and E. E. Buckley, Eds.) Vol. 3 New York: Academic Press 1971

Russell, F. E.: Marine Toxins and Venomous and Poisonous Marine Animals T. F. H. Publications, Neptune City, N. J. USA 1971 (new reprint in preparation)

Tsuda, K.: Über Tetrodotoxin, Giftstoff der Bowlfische. Naturwissenschaften *53*, 171 (1966)

6 Amphibia (Amphibians)

The class *Amphibia* contains about 2600 species; they are zoologically divided in Anura (amphibia without tails, e. g. frogs and toads) and Urodela (amphibia with tails, e. g. salamanders and newts); both of these orders are subdivided into numerous genera which can be widely differentiated by their morphology.

The amphibia must also be considered among the venomous animals, although they use their toxins for protection and not for attack. The venoms are produced in skin glands distributed over the whole surface of the body and secreted in small but steady amounts. Until recently it was commonly held that these secretions are used only against the natural predators. However, this is not the case. Investigations in the laboratory of the author showed without any doubt that these toxins primarily protect against microorganisms. The skin of amphibians would be a perfect substrate for bacteria and fungi, since it must be continuously moist to permit the exchange of oxygen and carbon dioxide. The moisture is provided by a mucous secreted from glands in the skin. On the other hand the amphibians live in an environment rich in microorganisms. We indeed found that detoxified animals died from skin infections within a few days, an observation that was independently made by H. Michl in Vienna. A thorough investigation of the antibiotic activity showed that the toxins were able to inhibit the growth of bacteria and fungi in concentrations as low as 10^{-3} to 10^{-5} mole per liter. Electron microscopy shows widespread damage within the cells including lesion of the cytoplasmic membrane, damage to the mitochondria (in fungi) and lesion of the ribosomes.

Poisoning from amphibians has practically never been observed; hence clinical syndrome and treatment need not be discussed here.

Distribution

The distributions of some of the most important Amphibian families are shown in Figs. 13–18.

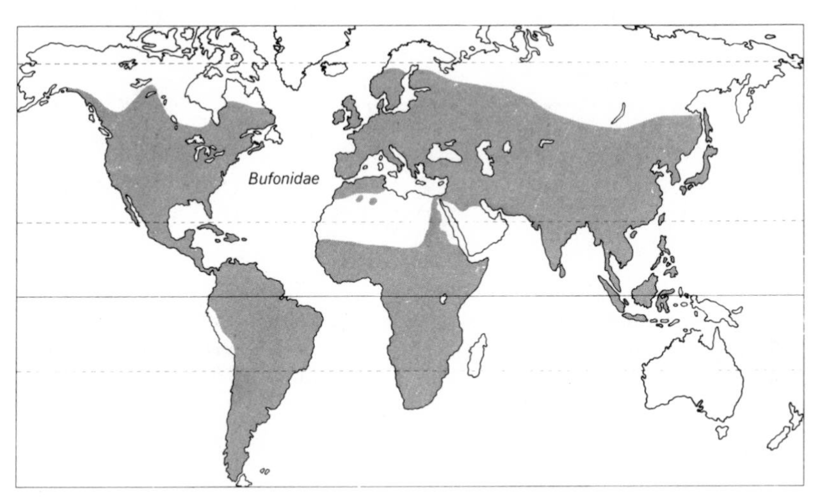

Fig. 13. Distribution of the bufonids (after Savage, 1973, from INTRODUCTION TO HERPETOLOGY, Third Edition, by Coleman J. Goin, Olive B. Goin and George R. Zug. W. H. Freeman and Company. Copyright © 1978)

Fig. 14. Distribution of the dendrobatids (after Savage, 1973, from INTRODUCTION TO HERPETOLOGY, Third Edition, by Coleman J. Goin, Olive B. Goin and George R. Zug. W. H. Freeman and Company. Copyright © 1978)

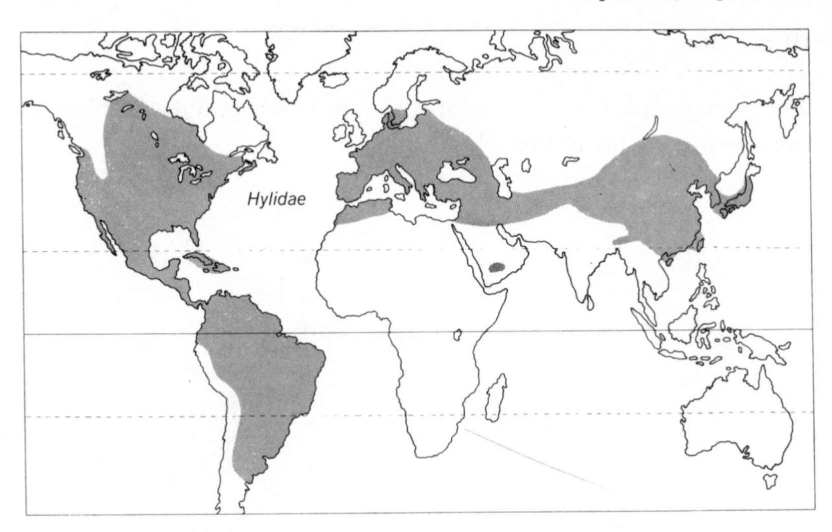

Fig. 15. Distribution of the hylids (after Savage, 1973, from INTRODUCTION TO HERPETOLOGY, Third Edition, by Coleman J. Goin, Olive B. Goin and George R. Zug. W. H. Freeman and Company. Copyright © 1978)

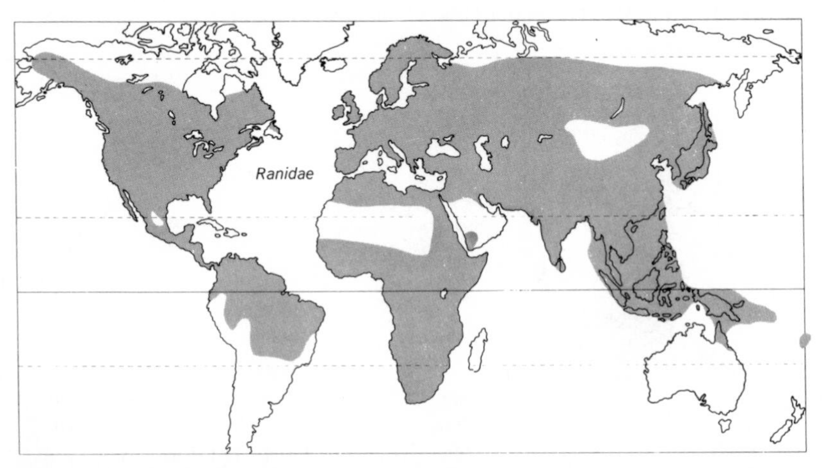

Fig. 16. Distribution of the ranids (after Savage, 1973, from INTRODUCTION TO HERPETOLOGY, Third Edition, by Coleman J. Goin, Olive B. Goin and George R. Zug. W. H. Freeman and Company. Copyright © 1978)

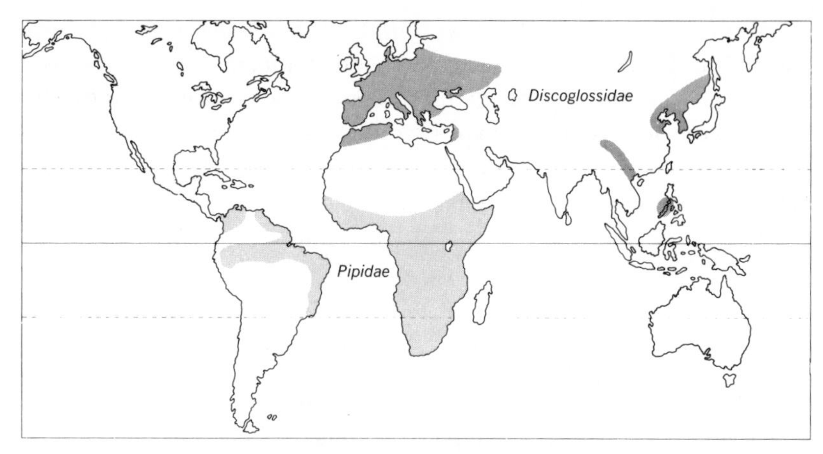

Fig. 17. Distribution of the discoglossids and pipids (after Savage, 1973, from INTRODUCTION TO HERPETOLOGY, Third Edition, by Coleman J. Goin, Olive B. Goin and George R. Zug. W. H. Freeman and Company. Copyright © 1978)

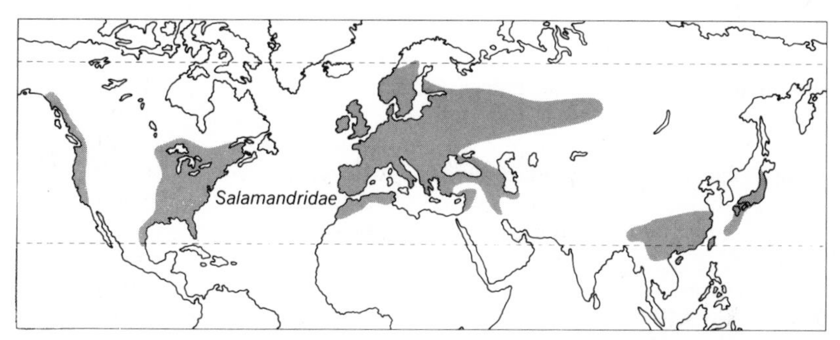

Fig. 18. Distribution of the salamandrids (after Savage, 1973, from INTRODUCTION TO HERPETOLOGY, Third Eition, by Coleman J. Goin, Olive B. Goin and George R. Zug. W. H. Freeman and Company. Copyright © 1978)

Chemistry

The chemistry of amphibian toxins has been the subject of many investigations throughout the last decade and is thus relatively well known. The crude secretions contain an amazing variety of different substances: simple biogenic amines, peptides, steroids and alkaloids. Pharmacologically they can be cardiotoxic, myotoxic, neurotoxic and hemotoxic, cholinomimetic as well as sympathomimetic, vasoconstrictive and hypotensive and locally anesthetic. Even one of the most potent hallucinogens, o-methyl-bufotenin is found. Some of these compounds belong among the most potent toxins yet known. Examples are listed in Table 33.

Table 33. Toxicity and effects of amphibian venoms

Compound	Distribution	Effect	LD_{50}, s. c. (γ/kg mouse)
Batrachotoxin	Phyllo bates aurotaenia	Cardio- and neurotoxic	2
Tetrodotoxin	Taricha toroba	Neurotoxic	8
Samandarin	Salamandra spp.	Central convulsion venom	1500
Bufotoxin	Bufo spp.	Cardiotoxic	400
Toxins from	Triturus spp.	Hemolytic and muscle-nerve-activity	6700 20000 1800
Pumiliotoxin A	Dendrobates spp.	Nerve-muscle-activity	2500
Pumiliotoxin B	Dendrobates spp.		1500
Pumiliotoxin C	Dendrobates spp.		1200
Serotonin	Hyla spp.	Vasoconstrictor	300
Dehydrobufotenin	Leptodactylus spp.	Convulsion venom	6000
O-methy-bufotenin	Bufo spp.	Halluscinogen	75000
Norepinephrin	Bufo spp.	Hypertensive	5000
Candicin	Bufo spp.	Cholinomimetic	10000
Leptodactylin	Leptodactylus spp.		10000
Zetekitoxin AB	Atelopus spp.	Convulsion venom, cardiotoxic,	11

6.1 Anura (Frogs, Toads)

6.1.1 Bufonidae (Common Toads)

Our knowledge of the toxicity of the toads dates back to early times. In ancient Chinese and Japanese medicine – 3000 years ago – a dried and powdered secretion (ch'an-su, Sen-so) was used as heart drug. It may still be found there (East and Southeast Asia) in the local pharmacies as a remedy for dropsy. Such substances were introduced in Europe in the 17th century and used for about 200 years until digitalis glycosides came into for the same purpose.

It is interesting to note that according to medieval superstition toads attain their toxicity from the Christrose, *Helleborus niger,* under which they frequently may be found. And in fact, hellebrigenin and bufotalidin are chemically identical compounds.

The substances from toad venom may be divided into three major groups.

A. Biogenic amines

These compounds are partly derivatives of brenzcatechin (catecholamine), partly indolalkylamines. The most important compounds are the following ones:

1. *Adrenaline* and *nor-adrenaline:* Adrenaline is well known from the animal kindom as a suprarenal hormone. It is also widely distributed in the skin gland secretions of amphibians as is nor-adrenaline, a sympathomimetic. Macht and Abel were the first to detect these substances in toad venom. Other substances isolated from toad secretions are dopamine and epinine (N-methyl-dopamine):

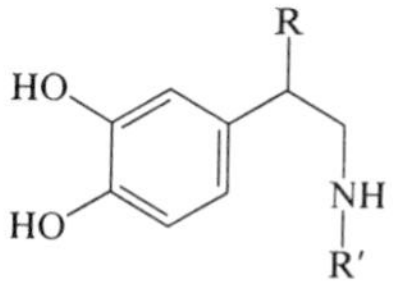

Adrenaline: R = OH, R' = CH_3
Nor-adrenaline: R = OH, R' = H
Epinin: R = H, R' = CH_3

The biosynthesis of these compounds starts from phenylalanine, which is oxidized via tyrosine to yield dioxphenylalanine. Decarboxylation then leads to dopamine, which is either methylated to epinin or hydroxylated to nor-adrenaline; methylation of nor-adrenaline yields adrenaline.

2. *Indolalkylamines:* The evidence for basic compounds with pharmacological activity in the skin gland secretion of toads was first provided by Phisalix, Bertrand and Handovsky. The structure of the first of these substances was elucidated by H. Wieland et al.:

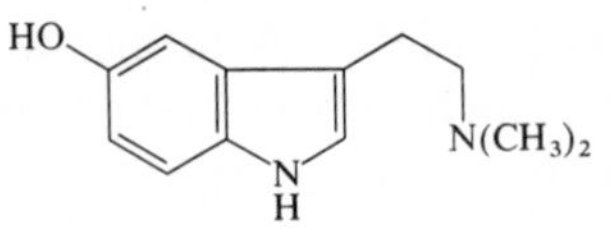

HO, $N(CH_3)_2$, N H — Bufotenin

O-methyl-bufotenin *(Bufo alvarius)* is one of the most potent hallucinogens. Bufotenin-containing drugs (from mushrooms) are used by South American Indians. Bufotenidin, which is the methylbetaine of bufotenin, possesses like all indolalkylamines a vasoconstrictive and hypotensive activity.

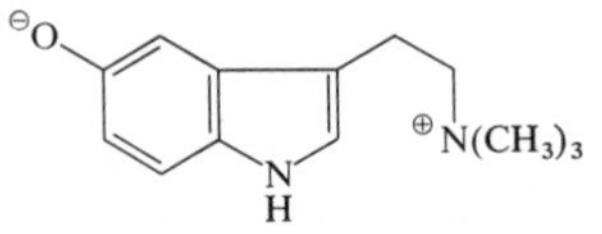

$^{\ominus}O$, $^{\oplus}N(CH_3)_3$, N H — Bufotenidin

Bufoviridin has been isolated from the skin of *Bufo viridis viridis.* Dilute hydrochloric acid cleavage yields bufotenin and sulfuric acid. The other idolalkylamines, bufothionin and dehydrobufotenin, are closely related to one another.

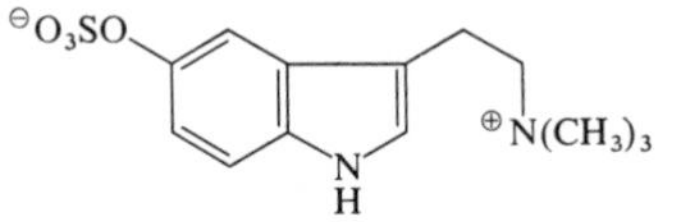

$^{\ominus}O_3SO$, $^{\oplus}N(CH_3)_3$, N H — Bufoviridin

H_3C, CH_3, $N^{\oplus}$, $^{\ominus}O$, N H — Bufothionin

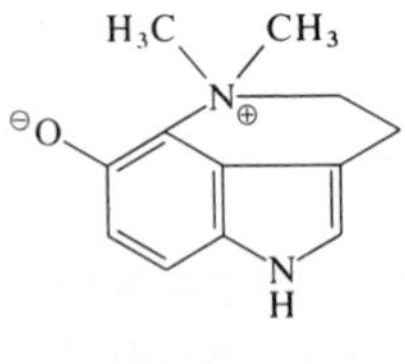

H_3C, CH_3, $N^{\oplus}$, $^{\ominus}O_3SO$, N H — Dehydrobufotenin

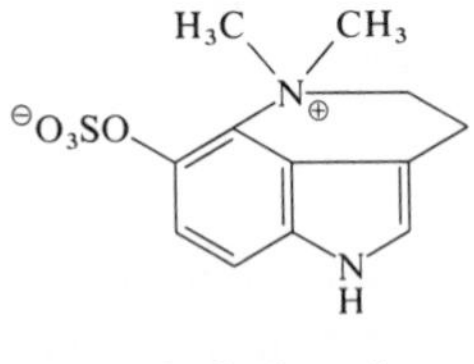

Bufothionin has been isolated from the skin gland secretion of *Bufo formosus* and *Bufo arenarum.* On reaction with warm, dilute hydrochloric acid dehydrobufotenin hydrochloride is formed. Dehydrobufotenin itself could be isolated from the secretion of *Bufo marinus.*

B. Bufogenines

The first investigations on the bufogenines were done by Faust in 1902. He assumed a close relationship between bufotalin isolated by him and the bile acids. Wieland et al. were able to obtain important details on the structure of the molecule; the final structure determination was achieved by K. Meyer in 1949. All of the bufogenines possess the same basic structure with cis-connection of rings A and B as well as C and D, and an unsaturated lactone ring (bufadienolid ring).

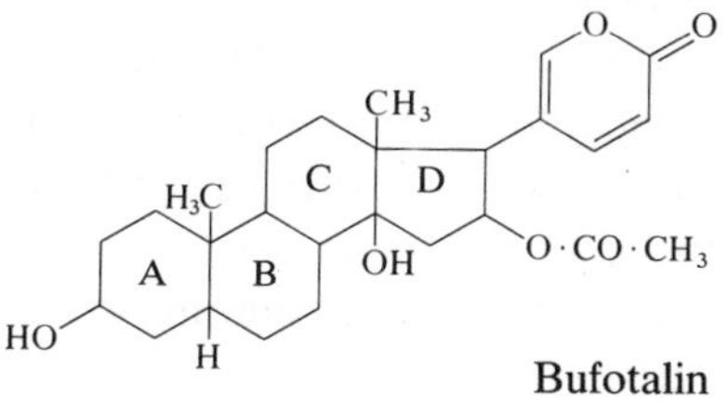

Bufotalin

All bufogenines act on the heart, generally as cardiotoxins.

C. Bufotoxins

The bufogenines are accompanied by the so-called bufotoxins. They are an independent class of compounds in that all of them are esters of suberylarginine with the corresponding bufogenines. Cleavage into these two components can only be done enzymatically, because under acidic conditions one mole of water is eliminated additionally. The elucidation of the structure was first done by H. Wieland and later on by K. Meyer. Biogenesis of bufogenines and bufotoxins starts from cholesterol.

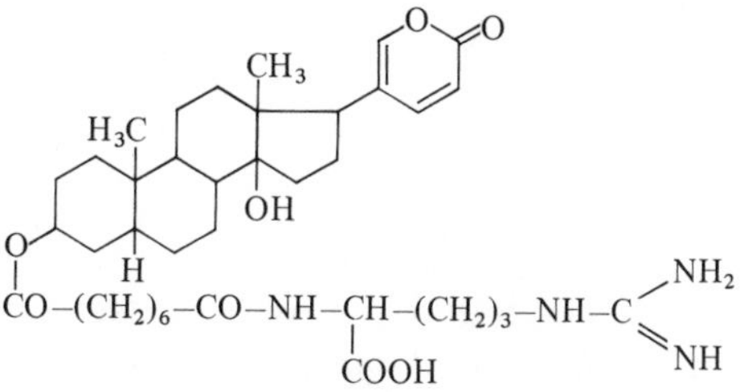

Bufotoxin

The physiologic activities of bufogenines and bufotoxins closely resemble those of digitalis toxins. They possess a positive inotropic effect; that is they increase the strength of contraction of (weak) heart muscle, they cause an increase of the tonus and a decrease of frequency.

Despite their use in East Asia, the bufogenines do not play a role in modern therapy, since drugs from digitalis plants are more readily available and possess fewer side effects.

The strong local anesthetic action of many bufogenines and bufotoxins is remarkable; it exceeds by far that of cocaine.

Many attempts were made to establish relationships between structure and toxicity. Indeed, a number were found. The unsaturated lactone ring is essential for the heart activity. Hydrogenation leads to inactivation. Furthermore, the stereochemistry of the basic structure is important. Absence of the hydroxy group at C-14 or its conversion into a 14.15 – oxido-group decreases the action on the heart. Instead, spasmodic activities arise.

6.1.2 Leptodactylinae

Leptodactylus pentadactylus and *L. ocellatus,* native to South and Central America, contain considerable amounts of 5-hydroxy-tryptamin derivatives (up to 10 mg per g skin): N-methyl-serotonin (5-hydroxy-N-methyl-tryptamin), bufotenin and bufotenidin. Besides those indolalkylamines, a phenolbetain, leptodactylin, is found:

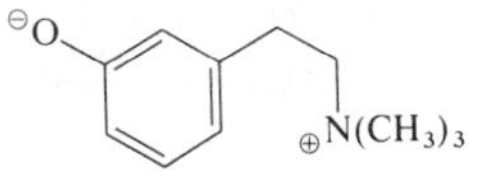

$^{\ominus}O$ $^{\oplus}N(CH_3)_3$
Leptodactylin

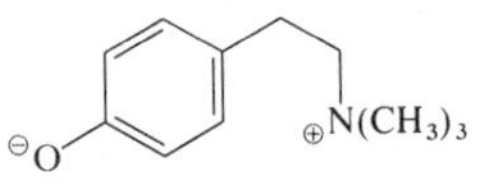

$^{\ominus}O$ $^{\oplus}N(CH_3)_3$
Candicin

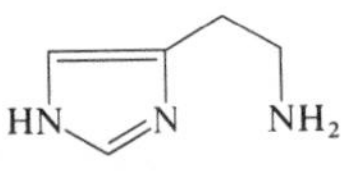

HN N NH_2
Histamine

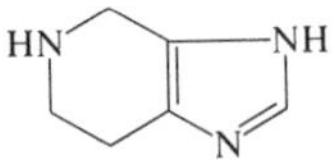

HN NH N
Spinaceamine

Its p-isomer, candicin, has also been found. In vivo leptodactylin is formed from m-tyrosine. The pharmacologic activity is manifested by a strong nicotine-like stimulation of the autonomic ganglia as well as the neuromuscular junction and finally a neuromuscular block. The physiological significance is still unclear.

Bufotenidin, dehydrobufotenin, leptodactylin, histamine and peptide similar to caerulin have been isolated from the skin of *L. vilarsi,* which is native to Ecuador. The skin of *L. pentadactylus labyrinthicus* contains histamine, N-methyl-histamine, N-acetyl-histamine and N. N.-dimethyl-histamine as well as spinaceamine, an imidazolylalkylamine. The latter can be synthesized in vitro by a biomimetic reaction of histamine and formalydehyde in aqueous solution at pH 6.8 and room temperature.

6.1.3 Dendrobatidae (Poison-dart Frogs, Coloured Frogs)

Dendrobatid toxins are completely different from the compounds discussed so far. They are alkaloids, more complicated, basic compounds. Dendrobatid frogs are long known to be venomous. The skin secretions are used by the Noanama, Choco and Cuna indians of Columbia as dart poisons.

These secretions and their components have been very thoroughly studied in recent years; in all, more than 100 species and populations of *Phyllobates spp.* and *Dendrobates spp.* have been investigated.

The structures of the toxins of *Phyllobates aurotaenia* (earlier names: *P. chocoensis, Dendrobates tinctorius, P. bicolor, P. latinosus)* have been established by Witkop et al. Four extremely toxic substances habe been isolated. They are steroid alkaloids, that do not belong to anyone of the steroid alkaloid types previously known. The first one is *batrachotoxin* (BTX), which is the most toxic component. It is the 20-ester of 2.4-dimethylpyrrol-3-carboxylic acid, with *batrachotoxinin* A, the second substance. The third alkaloid is *homobatrachotoxin* (formerly designated as *isobatrachotoxin),* the 20-ester of 2-ethyl-4-methyl-pyrrol-3-carboxylic acid and batrachotoxinin A. Finally the secretion contains a fourth, extremely instable substance, pseudobatrachotoxin, which on standing at room temperature decomposes to yield batrachotoxinin A.

The structure of these compounds was determined from x-ray crystallographic analysis of the p-bromobenzoate of batrachotoxinin A and from chemical data. The biological activity is highly

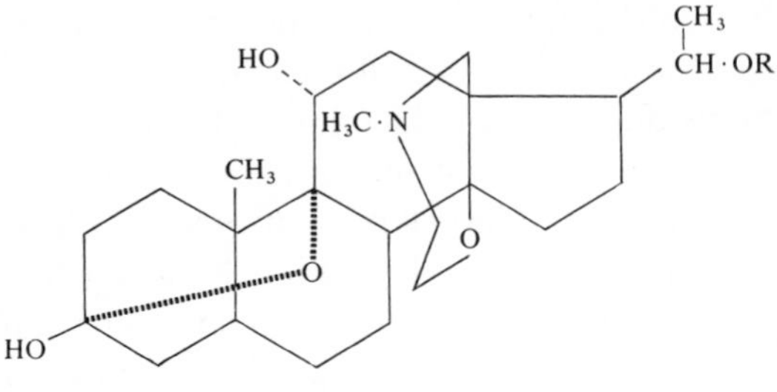

Batrachotoxinin A

Batrachotoxin R = —CO—

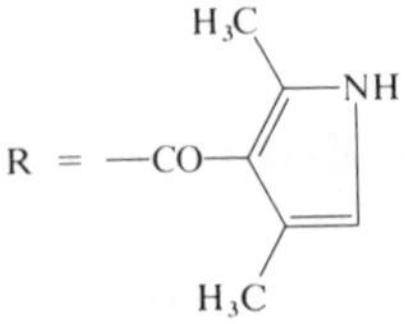

Homobatrachotoxin R = —CO—

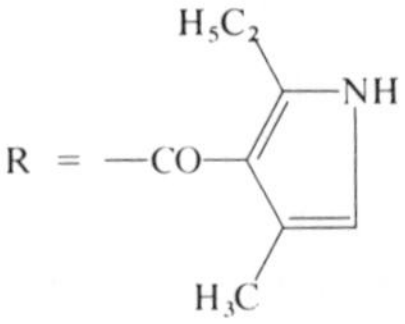

dependent on the nitrogen containing 7-membered ring, the heterocyclic system attached to C-20 and the 3β-hydroxy-3.9-oxido-hemiacetal group. On cleavage of the hemiacetal (e.g. by reduction with $NaBH_4$) the toxicity decreases to 1/100 of the original value. The physiological activity of batrachotoxin is directed at the central nervous system. In the nerve-muscle preparation it causes an irreversible block of the motor end-plates. The action comes from an irreversible increase of the permeability of the cell membranes for sodium ions without any disturbance of the potassium transport. The antagonist of batrachotoxin is tetrodotoxin (cf. p. 103). The symptoms of intoxication are muscle and finally respiratory paralysis; they differ significantly, however, from those of curare. Death occurs within a few seconds. An antidote is not yet known; not even to the indians of Columbia. Among the other *Dendrobatidae*, only *Phyllobates vittatus* (from Costa Rica) and *Ph. lugubris* (Panama) contain batrachotoxin, homobatrachotoxin and batrachotoxinin A.

Three alkaloids have been isolated from the skin of *Dendrobates pumilio* (Panama):

Pumiliotoxin A, $C_{19}H_{33}NO_2$
Pumiliotoxin B, $C_{19}H_{33}NO_3$
Pumiliotoxin C, $C_{13}H_{25}N$

Each frog contains about 0.2 mg of toxin. The isolation is extremely difficult because of the instability of the pumiliotoxins. The structure of PTX-C has been determined by X-ray crystallographic analysis. The type of structure was previously unknown in nature. PTX-A, PTX-B and PTX-C have also been isolated from *Dendrobates auratus,* together with other alkaloids of a closely related type. Subcutaneous injection of pumiliotoxins leads to death after strong clonic cramps.

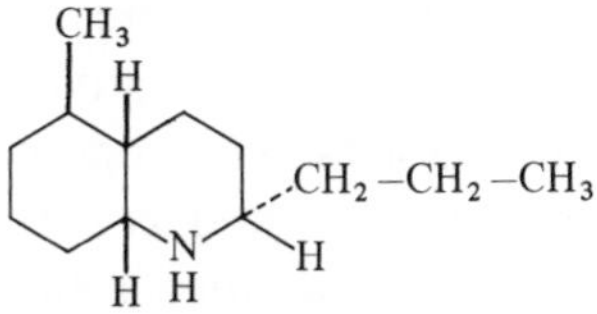

Pumiliotoxin C

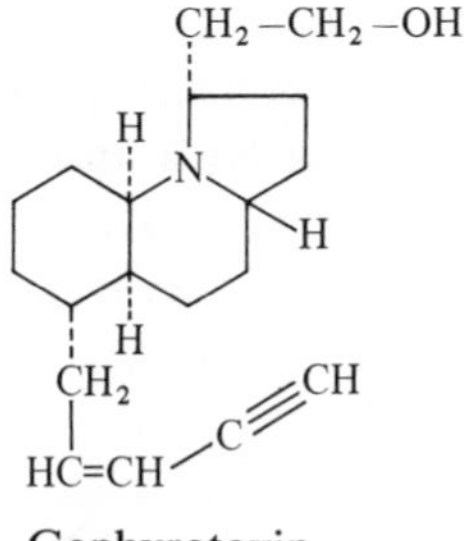

Gephyrotoxin

Alkaloids of another type have been isolated from *D. histrionicus;* they are spiropiperidine alkaloids:

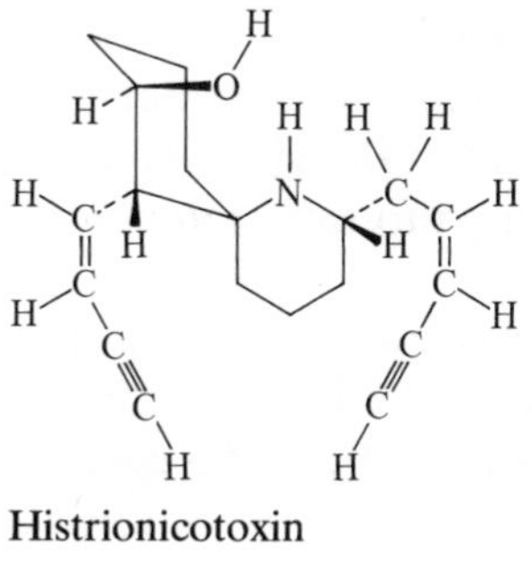

Histrionicotoxin

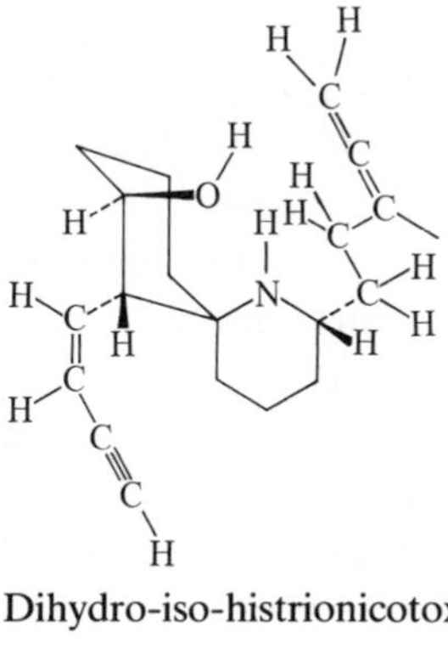

Dihydro-iso-histrionicotoxin

Other alkaloids with longer side chains have also been found; their structure, however, is not yet finally established.

Still another type of alkaloid could be isolated from *D. histrionicus,* the gephyrotoxins.

Those are the first substances found in the animal kingdom to possess allenic or even C-C-triple bonds. In addition to these unsaturated compounds, the partially or completely hydrogenated substances have been found.

6.1.4 Atelopodidae

Strong, dialyzable toxins have been isolated from the skin of *Atelopus species.* The zetekitoxins are especially abundant in *Atelopus zeteki,* the "Golden Arrowfrog" of Panama, in smaller amounts in *A. varius, A. varius ambulatorius, A. cruciger* and *A. planispima.* Secretions of these frogs have been used by the natives of Panama, Costa Rica and Columbia as dart poisons. The venom has a strongly hypotensive action, causes convulsions and is cardiotoxic. Chemical investigations are still incomplete. From the results obtained so far it is known that the substances are readily soluble in water; they can be precipitated by acetone from aqueous solutions. The zetekitoxins AB and C could be isolated in pure forms. The structures, however, are not yet clear. They are no peptides, carbohydrates or steroids. A guanidine group has been detected, and thus a certain similarity with tetrodotoxin seems to exist in this regard, but both compounds possess different pharmacological activities.

The lethal dose for zetekitoxin AB is 11 μg/kg, for zetekitoxin C 80 μg/kg (mouse, s. c.), both of them being hypotensive. On the other hand tetrodotoxin has been isolated from *A. varius ambulatorius* and *A. varius varius.* 30% tetrodotoxin together with another toxin, chiriquitoxin, have been found in the skin glands of *A. chiriquensis.* Chiriquitoxin is closely related to tetrodotoxin; its structure, however, is not yet clear.

Tetrodotoxin and chiriquitoxin occur not only in the skin but also in the egg clusters, where they exist in bound form. This follows from the observation that they can be extracted from the clusters only by use of 3% acetic acid, even though they are otherwise readily water soluble. This is most important in explaining why these frogs as well as other tetrodotoxin containing species are able to tolerate this extremely potent toxin in considerable amounts.

6.1.5 Hylidae (Leaf Frogs)

Hyla arborea L. and subspecies are distributed all across Central and Southern Europe, North Africa, the Caucasus and the Urals through Asia to Japan. The skin secretion contains a hemolytic peptide of unknown structure that is active even in dilutions of 1:200,000. Serotonin occurs in practically all of the hylidae. *H. caerula* contains histamine, serotonin and caeruleine, a strong hypotensive peptide of the following structure:

Pyroglutamyl-Glu-Asp-Tyr-(SO_3H)-Thr-Gly-Tryp-Met-Asp-Phe-NH_2

Bufotenin has been found in *H. pearsoniana* and *H. peroni.* A great number of other *Hyla* and *Trachycephalus* species have been investigated pharmacologically or toxicologically; little or nothing is known as yet about their chemistry.

6.1.6 Phyllomedusae (Grip Frogs)

The genus *Phyllomedusa,* native to South America, is closely related to the *hylidae* and also contains low molecular weight peptides in their skin glands. A bradykinin-type peptide, phyllokinin, has been isolated from *Phyllomedusa rohdei;* it has the following sequence:

Pyroglutamyl-Asp-Pro-Asp-Arg-Phe-Ile-Gly-Leu-Met-NH_2

6.1.7 Ranidae (True Frogs)

The family of *Ranidae* has been investigated very thoroughly. The majority of the species contain indolalkylamines known from the toads, especially serotonin.

In *Rana temporaria* and *R. nigromaculata,* bradykinin, a hypotensive peptide, has been found:

Arg-Pro-Pro-Gly-Phe-Ser-Pro-Phe-Arg

Bradykinin also acts on the smooth muscles.

The skin gland secretion of the water frog *Rana esculenta,* distributed all over Europe, contains four peptides besides a large number of free amino acids. Aqueous solutions of the secretion are toxic even in high dilution. The heart muscle of a frog is paralyzed in concentrations of only 1:500000. The LD_{50} of the dried crude secretion has been determined to be 6–12 mg/kg (rabbit). Moreover a strong hemolytic protein has been isolated.

6.1.8 Discoglossidae (Disc Tongues)

Among the *Discoglossidae,* particularly the red unk, *Bombina bombina,* and the yellow unk, *B. variegata,* have been investigated by H. Michl et al. The skin secretions have a characteristic smell; they cause severe sneezing and catarrh-like symptoms in humans. The dried skin secretion of *B. bombina* contains 10% serotonin, as well as free amino acids and four basic peptides, the structure of which are as follows:

Ala-Glu-His-Phe-Ala-Asp $(NH_2)_2$

The secretion of *B. variegata* consists of 12 α-amino acids, γ-amino-butyric acid and 5-hydroxy-tryptamine as well as two nonapeptides of the following structure:

Ser-Ala-Lys-Gly-Leu-Ala-Glu-His-Phe and
Gly-Ala-Lys-Gly-Leu-Ala-Glu-His-Phe

Moreover a hemolytic polypeptide of a molecular weight of 87,000 Daltons has been found. It consists of two subunits; the activity is dependent on the presence of calcium ions.

6.1.9 Pipidae (Tongue-less Frogs)

Xenopus laevis is the only species of the family *Pipidae* investigated so far. Serotonin and bufotenidin have been found in the skin secretion.

6.2 Urodela (Newts, Salamanders)

Within the order *Urodela,* the genera *Salamandra* and *Triturus* have been especially well studied.

Salamanders, particularly the European Fire Salamander, since ancient times have been known to be toxic. Like the toads, the salamanders played a role in mythology for millennia. In ancient Persian mythology, it was the animal capable of extinguishing fire, and in medieval times it was one of the symbols of alchemists trying to convert lead into gold. After initial investigations by *Faust, Zalesky* and *Gessner,* the skin gland secretion was studied by *Schöpf* and *Habermehl.* A series of papers described the isolation and purification of the toxins. Studies on the structures of these compounds showed them to be steroid alkaloids. The structures were finally determined by X-ray crystallography. These alkaloids may be differentiated into three groups according to structure: those with an oxazolidine system, with a carbinolamine system and with neither.

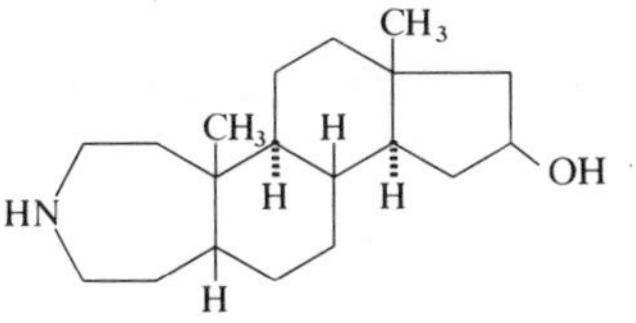

Samanin

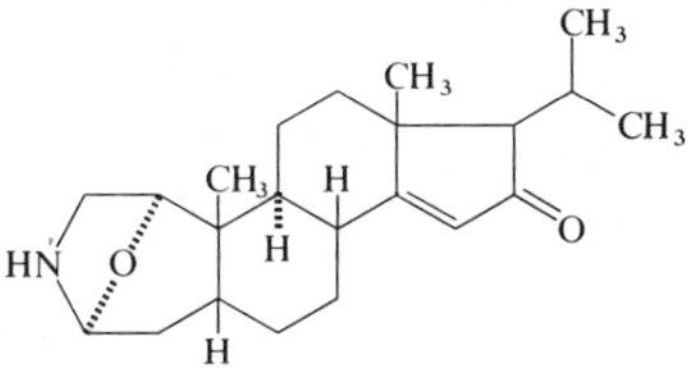

Samandenon

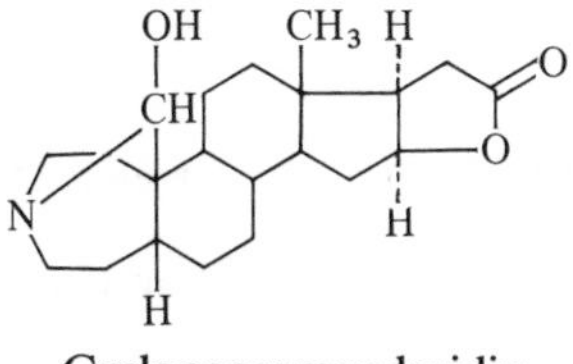

Cycloneosamandaridin

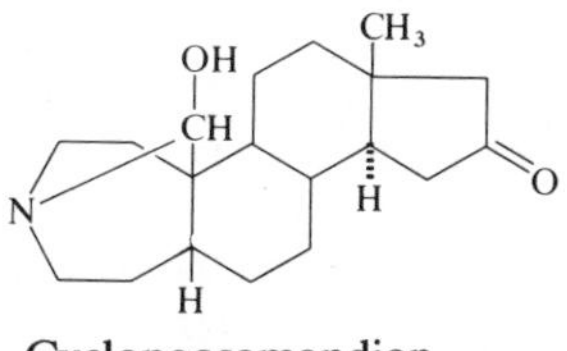

Cycloneosamandion

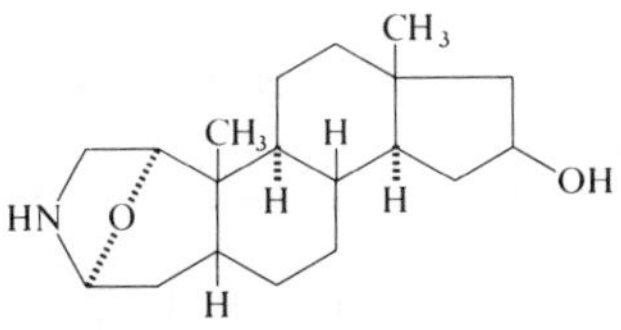

Samandarin

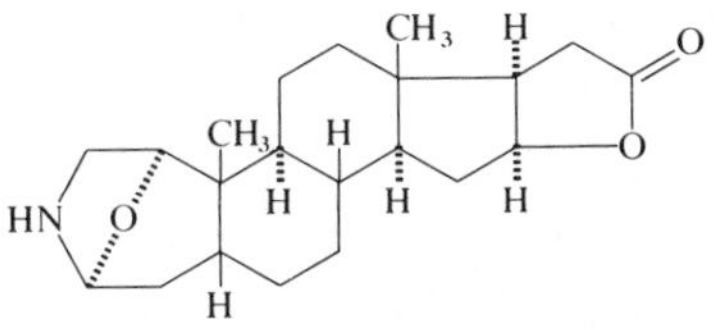

Samandaridin

Samandarin is a very potent neurotoxin that acts on the central nervous system and causes convulsions. Death results from primary respiratory paralysis. Samandarin, moreover, possesses remarkable hypertensive and local anesthetic properties. The European newts *T. vulgaris, T. cristatus, T. alpestris* and *T. marmoratus* (genus *Triturus)* have been studied, mainly pharmacologically. The extremely strong hemolytic activity of the secretions up to a concentration of $1:10^9$ is most striking. In addition amylases, phosphatases and arylamidases have been found, but no steroids or alkaloids.

Tarichatoxin has been isolated from the Californian newts *Taricha torosa, T. rivularis* and *T. granulosa.* It is identical to tetrodotoxin derived from the buffer fish *Sphoeroides rubripes* and others (see p. 103). The structure has been determined by chemical degradation and finally by X-ray crystallographic analysis of the tetrodonic acid hydrobromide. The occurrence of tetrodotoxin in newts of the genus *Triturus* is still debatable.

Phylogenetically this toxin is of special interest because it occurs in octopuses, fishes, newts and frogs (but in a few frog genera only). The biological significance is completely uncertain. According to what we presently know, it is not used to protect against microorganisms like many others of those substances.

References

Bücherl, W., Buckley, E. E. (Eds.):Venomous Animals and Their Venoms, Vol. II New York: Academic Press 1971

Habermehl, G.: Naturwissenschaften *56*,615 (1969)

Habermehl, G.: Naturwissenschaften *62*,15 (1975)

Tokuyama, T., Daly, J., Witkop, B., Karle, I. L., Karle, J.: J. Amer. Chem. Soc. *90*,1917 (1968)

Fuhrman, F. A., Fuhrman, G. J., Dull, D. L., Mosher, H. S.: Agricultural and Food Chemistry *17*,417 (1969)

Shindelman, J., Mosher, H. S., Fuhrman, F. A.: Toxicon *7*,315 (1969)

Tokuyama, T., Daly, J. W., Witkop, B.: J. Am. Chem. Soc. *91*,3931 (1969)

Witkop, B.: Experienta (Basel) *27*,1121 (1971)

Warnick, J. E., Albuquerque, E. X. Onur, R., Jansson, S. E., Daly, J. W., Witkop, B.: J. Pharm. Exp. Ther. *193*,232 (1974)

Lampa, A. J., Albuquerque, E. X., Sarvey, J. M., Daly, J. W., Witkop, B.: Exp. Neurology *47*,558 (1975)

Brown, G. B., Kim, Y. H., Mosher, H. S., Fuhrman, J., Fuhrman, F. A.: Toxicon *15*,115 (1977)

Pavelka, L. A., Kim, Y. H., Mosher, H. S.: Toxicon *15*,135 (1977)
Mar, A., Michl, H.: Toxicon *14*,191 (1977)
Daly, J. W., Witkop, B., Tokuyama, T., Nishikawa, T., Karle, I. L.: Helv. Chim. Acta *60*,1128 (1977)
Savage, I.: The Geographic Distribution of Frogs, in Vial, I. L. (Ed.) Evolutionary Biology of Anurans, Univ. of Missouri Press, 1973. Figs. taken from C. J. Croin, Olive B. Croin and G. R. Zug, Introduction to Herpetology, 3rd Edt., 1978, Freeman, S. Francisco.

7 Reptilia (Reptiles)

Except for the gila monsters, *Heloderma*, the snakes are the only venomous animals of this class. Among the some 2,000 species of snakes about 400 species are toxic. They are classified into four families:

1. *Elapidae* (cobras, kraits, coral snakes, mambas)
2. *Hydrophiidae* (sea snakes)
3. *Viperidae* (vipers)
4. *Crotalidae* (rattlesnakes, pit vipers)

Another 400 species for the family *Colubridae* (which contains a total of 1,500) must be considered venomous. Even though only a few are known to have been responsible for a limited number of poisonings, it would be erroneous to believe that just these few species are toxic.

This classification follows the order generally used in zoology. The reader should note, however, that the genera *Agkistrodon, Bothrops, Crotalus, Sistrurus* and *Lachesis,* attributed to the *Crotalidae,* occasionally have been counted among the vipers.

Distribution

The snakes are among the most widely distributed animals. They inhabit all tropical, subtropical and most of the temperate zones of the earth. A difference, however, can be observed in the distribution of individual families. As a result of this broad distribution and their relatively great numbers, there are more accidents involving snakes than any other kind of venomous animals. Snakes possess a complete venom apparatus, consisting of highly specialized glands (parotid gland, Duvernoy's gland) and venom ducts draining into the fangs (except for the aglyphic snakes, which do not possess these fangs).

Snake venom is used to kill or paralyze prey, but it also facilitates digestion of the prey, which is swallowed whole. Snakes are more timid than aggressive towards humans. They only bite if they are threatened, frightened or stepped upon when hidden under grass or leaves.

Poisoning

Estimates of the number of envenomations vary greatly. According to thorough and rather reliable statistics based on hospitalized cases only (but not taking into account the USSR, China and Europe), there are 40,000 fatalities yearly; nearly 50% occur on the Indian subcontinent. If one assumes a mortality rate of 2.4% for snake bites, which is reasonable, 1.7 million should occur annually. In reality this number is much higher due to undocumented cases. There are unofficial estimates for the Indian subcontinent (comprising India, Bangla Desh, Pakistan and adjoining areas) of approximately 100,000 fatalities per year.

Very interesting and impressive data are given in Table 34:

Table 34. Severity of snake bites

Degree of envenomation	USA	India	Malaya	Natal
Mild	64%	30%	53%	88%
Severe	36%	40%	46%	10%
Fatal	0,2%	30%	1.5%	2%

In the Hospital Vital Brazil, a special clinic in Sao Paulo for the treatment of accidents involving venomous animals 15,709 cases were treated in the years 1954–1965. Among those, 1,323 bites were non-venomous and 1,718 from venomous snakes; 30 of the latter fatal. In the USA about 45,000 snakebite cases per year are recorded, among them 8,000 from venomous snakes. As a result of the excellent methods of treatment developed in recent years, there are fewer than 15 fatalities annually.

The mortality rates differ considerably according to the species (cf. Table 35).

Table 35. Mortality depending on species

Species	Severe injury	Mortality rate
All venomous snakes	24%	2.4%
Elapidae		25%
Viperidae		2%
Crotalidae		2.3%
Dendroaspis polylepis		100%
Naja naja		32%
Bitis lachesis	27%	5.2%
Vipera palaestinae	28%	6.6%
Vipera berus	5%	1%
Echis carinatus	30%	20%
Crotalus durissus terrificus		12%
Bothrops jararaca		0.3%
Bothrops jararacussu		7.2%
Lachesis muta		0.2%
Agkistrodon microlepidota		20%

Clinical Symptoms

The clinical symptoms of a snakebite depend on three factors:

1. On the amount of the venom injected. In many snakes, especially vipers and elapids, it is possible to distinguish between defensive and predatory bites. In a defensive bite relatively little venom is ejected, but just the opposite is true in predation. Fortunately most bites come from self-defense, but even those small amounts may cause severe injury.

Examples of the danger posed by venoms are given in Table 36.

Most rattlesnakes discharge between 25 and 75% of their venom in bites inflicted on humans; vipers discharge slightly less. There appears to be a greater variation in the amount an elapid may discharge. Many victims of elapid venom poisoning show minimal signs and symptoms; others may die. On the other hand sea snakes usually discharge most of their available venom.

2. The symptoms depend on the site of the bite: Venom injected into a muscle produces symptoms less dangerous than if a blood vessel had been hit, in which case the chance of fatality is rather high due to the very fast transport of cardio- and neurotoxins to heart

Table 36. Lowest lethal dose of some snake venoms

Species	Injected amount venom/bite	Lethal dose in human (75 kg)
Elapidae		
Naja naja (cobra)	210 mg	15 mg
Naja bungarus (royal cobra)	100 mg	12 mg
Bungarus candidus (krait)	5 mg	1 mg
Bungarus caeruleus (krait)	10 mg	6 mg
Dendroaspis polylepis (black mamba)	1000 mg	120 mg
Viperidae		
Vipera russellii (Russellviper)	70 mg	42 mg
Echis carinatus	12 mg	5 mg
Vipera berus (common viper)	10 mg	75 mg
Crotalidae		
Bothrops neuwiedii (urutu)	200 mg	200 mg
Trimeresurus gramineus	14 mg	100 mg

Table 37. Clinical picture of a number of snake bites

Family	Local effects	Systemic effects
ELAPIDAE		
1) Cobras	Swelling; necrosis; various pains; local effects may be absent	Neurotoxic and cardiotoxic
2) Kraits	None	Neurotoxic
3) Micrurus	Pains; moreover no local symptoms	Neurotoxic
VIPERIDAE	Swelling; strong pains; necrosis very rare	Tissue destruction (hemorrhages; breakdown of coagulation; cardiovascular shock)
CROTALIDAE		
1) Crotalus	Strong pains changing to apathy	Neurotoxic
2) Bothrops	Strong pains	Proteolytic
3) Lachesis	Necrosis	Breakdown of coagulation
HYDROPHIIDAE	None	Myotoxic

and brain. Moreover, snake venoms contain substances that act on the clotting factors of blood, or can be hemolytically active.

3. The general physical condition of the victim is very important, since heart and/or circulatory distress are frequent. Primary shock and collapse are not rare occurrences; they are not necessarily a direct effect of the venom but may occur as a consequence of nervous failure.

Special symptoms will be discussed under the individual species; some general data, however, are given in Table 37.

Treatment

The best therapeutic measure against snakebite is the application of antivenin. With proper application it may be useful hours, even days after the bite. Some difficulties, however, result from side effects and adverse reactions. Antivenin not highly purified may cause adverse reactions in 30% of the cases treated; 0.3% of the cases may be fatal as a result of anaphylactic shock. 75% of all snakebites are not serious, and application of antivenin should be restricted to serious cases.

Many experimental investigations as well as clinical observations have shown that methods formerly recommended such as cauterization, suction, incision of the bite (Maimonides, 1135–1206) or injection of potassium permanganate or other chemicals are completely useless. Amazingly enough they have been used uncritically up to the present day and in many cases may cause more harm than the bite itself. The same is true for fasciotomy, recommended in recent years. In most cases immobilization of the afflicted limb, reassurance of the victim, application of analgesics and, if necessary, anti shock therapy suffice as first aid.

The use of tourniquets is still controversial. Applied improperly they can lead to disaster. It certainly should not be used by untrained persons; even physicians may have difficulty in distinguishing between the different kinds of vascular occlusion.

In any case the patient should be seen by a doctor or brought to the intensive care unit of a hospital. Antivenin should be injected into the immediate area of the bite; in serious cases, also intravenously. Beforehand, Opthalmo- and Intradermo- reactions are used to check for adverse reactions. Adrenalin should be prepared in a second syringe for emergency use. It is useful to know that children require the same amount of antivenin as adults.

Chemistry

Snake venoms have been extensively investigated from the chemical standpoint. These venoms can be obtained by "milking" the snakes. They are seized behind the head, the posterior part of it being held between the forefinger and the thumb; the other fingers hold the neck against the hand. The body of the snake is kept firmly under the upper arm. Then the mouth is opened and the fangs held over a beaker. After slight pressure on the parotids the venom drops from the fangs. The colorless or pale yellow liquid is dried in a desiccator; the dry weight is about 10%. The crude venom is stable in dried form for many years, whereas solutions are unstable even in the refrigerator. We know that the venomous properties come from specific toxins, peptides with 60–70 amino acids. Most of these are cardiotoxins or neurotoxins. One exception is the viperotoxin, a neurotoxin with 108 amino acids. The primary structure of many snake toxins has been determined.

In addition to the toxins, the crude venoms contain many enzymes, e. g. phospholipases, endopeptidases, exopeptidases, proteinases. They are also toxic in that they lead to a rapid decrease of blood pressure, change the clotting properties of blood or damage blood vessels and tissues.

Beyond their medical aspects, snake venoms have proven to be extraordinarily useful substances for neurophysiological and biochemical studies. Moreover they have led to important and interesting new findings in phylogeny. Intensive investigation has shown that the sea snake toxins, neurotoxic peptides with 62 amino acid residues, are the simplest toxins, and that the cardiotoxic peptides of 60 amino acids as well as the long-chain neurotoxins with 71–74 amino acids, as found in the elapids, have evolved from them.

Figure 19 is a diagram of the evolution of proteroglyphe toxins. In this regard it is interesting to take a look at the primary structure of the cardiotoxins and neurotoxins from the same species. Table 38 shows the primary structure of snake toxins as far as they are known at present.

The differences can be easily seen.

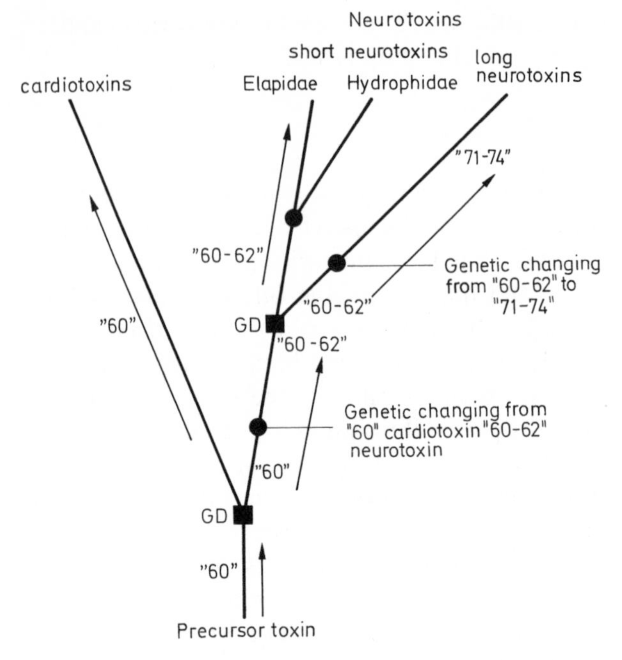

Fig. 19. Diagram of the evolution of snake toxins. The arrows show the direction of evolution; GD – gene duplication. The figures specify the number of amino acid residues of each toxin.

Tab. 38. Compilation of amino acid sequences of snake venom toxins (according to D. Mebs)

The sequences are grouped as follows:

1–31	– short neurotoxins (postsynaptic action)
32–56	– long neurotoxins (same action)
57–69	– so called angusticeps-type toxins of not yet defined pharmacological action
70–116	– cardiotoxins, cytotoxins, direct lytic factors (toxins active towards various kinds of cell membranes)
117–123	– phospholipases A (predominantly of presynaptic action) compared with pancreas enzyme
124–130	– protease inhibitors compared with pancreatic inhibitor and β_1-bungarotoxin, B-chain
131–133	– crotamine (affecting sodium channels) etc.

Abbreviations

N.n.a.	– Naja naja atra
N.haj.	– Naja haje
N.h.h.	– Naja haje haje
N.h.ann. (a.)	– Naja haje annulifera
N.melan.	– Naja melanoleuca
N.m.moss.	– Naja mossambica mossambica
N.nig.	– Naja nigricollis
N.niv.	– Naja nivea
N.n.sia.	– Naja naja siamensis
N.n.	– Naja naja
N.n.n.	– Naja naja naja
N.n.ox.	– Naja naja oxiana
O.hann.	– Ophiophagus hannah
H.haem.	– Hemachatus haemachatus
D.p.poly.	– Dendroaspis polylepis polylepis
D.j.kaim.	– Dendroaspis jamesonii kaimosae
D.vir.	– Dendroaspis viridis
D.ang.	– Dendroaspis angusticeps
B.m.	– Bungarus multicinctus
L.sem.	– Laticauda semifasciata
L.l.	– Laticauda laticaudata
L.c.	– Laticauda colubrina
E.sch.	– Enhydrina schistosa
H.cyan.	– Hydrophis cyanocinctus
L.hard.	– Lapemis hardwickii
P.pl.	– Pelamis platurus
Aip.l.	– Aipysurus laevis
Cr.d.t.	– Crotalus durissus terrificus
Cr.v.h.	– Crotalus viridis helleri
Cr.v.v.	– Crotalus viridis viridis
Not.s.s.	– Notechis scutatus scutatus

One-letter notation for amino acid sequences:

A – Ala, C – Cys, D – Asp, E – Glu, F – Phe, G – Gly, H – His, I – Ile, K – Lys, L – Leu, M – Met, N – Asn, P – Pro, Q – Gln, R – Arg, S – Ser, T – Thr, V – Val, W – Trp, Y – Tyr.

The separation into individual components can be achieved by means of modern chromatographic methods. Biological activities can then be determined, and finally the chemical structure elucidated.

1. N. n. a., Cobrotoxin	LECHNQQSSQTPTTTGCSGGETNCYKKRWRD		–H–––	RGYRTERGC	––GCPSV	–NIEINCCTT	–DRCNN
2. N. h. a., α-toxin	LECHNQQSSQPPTTKTCP	–GETNCYKKRWRD	–H–––	RGSITERGC	––KGIEINCCTT		–DKCNN
2a. N. h. h., CM-6							
2b. N. niv., δ-toxin							
3. N. nig., α-toxin	LECHNQQSSQPPTTKTCP	–GETNCYKKVWRD	–H–––	RGTIIERGC	––GCPTVK	–PGIKLNCCTT	–DKCNN
4. H. hem., toxin-II	LECHNQQSSQPPTTKSCP	–GDTNCYNKRWRD	–H–––	RGTIIERGC	––GCPTVK	–PGINLKCCTT	–DRCNN
5. H. hem., toxin-IV	LECHNQQSSQTPTTQTCP	–GETNCYKKQWSD	–H–––	RGSRTERGC	––GCPTVK	–PGIKLKCCTT	–DRCNK
6. N. n. ox., α-toxin	LECHNQQSSQPPTTKTCS	–GETNCYKKWWSD	–H–––	RGTIIERGC	––GCPKVK	–PGVNLNCCRT	–DRCNN
7. N. m. moss., neurotox. I	LECHNQQSSEPPTTTRCSGGETNCYKKRWRD		–H–––	RGYRTERGC	––GCPTVK	–KGIELNCCTT	–DRCNN
8. N. m. moss., neurotox. III	LNCHNQMSAQPPTTTRCSRWETNCYKKRWRD		–H–––	RGYKTERGC	––GCPTVK	–KGIQLHCCTS	–DNCNN
9. N. melanol., toxin d	MECHNQQSSQPPTTKTCP	–GETNCYKKQWSD	–H–––	RGTIIERGC	––GCPSVK	–KGVKINCCTT	–DRCNN
10. N. niv., toxin-β	MICHNQQSSQRPTIKTCP	–GETNCYKKRWRD	–H–––	RGTIIERGC	––GCPSVK	–KGVGIYCCKT	–DKCNR
11. N. h. h., CM-10a	MICHNQQSSQPPTIKTCP	–GETNCYKKQWRD	–H–––	RGTIIERGC	––GCPSVK	–CKGVGIYCCKT	–DKCNR
12. D. p. poly., toxin-α	RICYNHQSTTRATTKSCE	––ENSCYKKYWRD	–H–––	RGTIIERGC	––GCPKVK	–PGVGIHCCQS	–DKCNY
13. D. j. kaim., V_n^I1	RICYNHQSTTPATTKSC	––GENSCYKKTWSD	–H–––	RGTIIERGC	––GCPKVK	–QGIHLHCCQS	–DKCNN
14. D. vir., toxin 4.11.3	RICYNHQSTTPATTKSC	––GENSCYKKTWSD	–H–––	RGTIIERGC	––GCPKVK	–RGVHLHCCQS	–DKCNN
15. E. sch., toxin 4	MTCCNQQSSQPKTTTNC	–A–ESSCYKKTWSD	–H–––	RGTRIERGC	––GCPQVK	–PGIKLECCHT	–NECNN
15a. L. hard., neurotoxin							
16. P. pl., pelamitoxin a	MTCCNQQSSQPKTTTNC	–A–ESSCYKKTWSD	–H–––	RGTRIERGC	––GCPQVK	–SGIKLECCHT	–NECNN
17. E. sch., toxin 5	MTCCNQQSSQPKTTTNC	–A–ESSCYKKTWSD	–H–––	RGTRIERGC	––GCPQVK	–SGIKLECCHT	–NECNN
17a. H. cyan., hydrophitoxin b							
18. H. cyan., hydrophitoxin a	MTCCNQQSSQPKTTTNC	–A–ESSCYKKTWSD	–H–––	RGTRIERGC	––GCPQVK	–KGIKLECCHT	–NECNN
19. Aip. l., toxin a	LTCCNQQSSQPKTTTDC	–A–DNSCYKKTWQD	–H–––	RGTRIERGC	––GCPQVK	–PGIKLECCKT	–NECNN
20. Aip. l., toxin b	LTCCNQQSSQPKTTTDC	–A–DNSCYKMTWRD	–H–––	RGTRIERGC	––GCPQVK	–PGIKLECCKT	–NECNN
21. Aip. l., toxin c	LTCCNQQSSQPKTTTDC	–A–DNSCYKKTWKD	–H–––	RGTRIERGC	––GCPQVK	–PGIKLECCKT	–NECNN
22. N. h. a., CM-14	MICHNQQSSQPPTIKTCP	–GETNCYKKRWRD	–H–––	RGTIIERGC	––GCPSVK	–KGVGIYCCKT	–NKCNR
23. N. h. a., CM-10	MICYKQQSLQFPITTVCP	–GEKNCYKKQWSG	–H–––	RGTIIERGC	––GCPSVK	–KGIEINCCTT	–DKCNR
24. N. h. a., CM-12	MICYKQRSLQFPITTVCP	–GEKNCYKKQWSG	–H–––	RGTIIERGC	––GCPSVK	–KGIEINCCTT	–DKCNR
25. D. p. poly., FS 2	RICYNHQSTTPATTKSC	––GENSCYKKTWSD	–H–––	RGTIIERGC	––GCPKVK	–QGIHLHCCQS	–DKCNN
26. L. semifas. erabutoxin a	RICFNQHSSQPQTTKTCPSGESSCYNKQWSD		–F–––	RGTIIERGC	––GCPTVK	–PGIKLSCCES	–ECVNN
27. L. sem., erabutoxin b	RICFNQHSSQPQTTKTCPSGESSCYHKQWSD		–F–––	RGTIIERGC	––GCPTVK	–PGIKLSCCES	–EVCNN
28. L. sem., erabutoxin c	RICFNQHSSQPQTTKTCPSGESSCYHKQWSD		–F–––	RGTIIERGC	––GCPTVK	–PGINLSCCES	–EVCNN
29. L. c., L. l., Laticotox. a	RRCFNHPSSQPQTNKSCPPGENSCYNKQWRD		–H–––	RGTITERGC	––GCPTVK	–PGIKLTCCQS	–EDCNN
30. L. l., Laticotoxin b	RRCFNHPSSQPQTNKSCPPGENSCYNKQWRD		–H–––	RGTITERGC	––GCPQVK	–SGIKLTCCQS	–DDCNN
31. N. h. h., CM-2	FTCF–––TTPSDTSETCPDGQNICYEKRWNS		–H–––	QGVEIK	–GCVASCPEFESKRFYLLCCRI		–DNCNK

32. N. n. sia., toxin 3	IRCF−−−ITPDITSKDCPNG −HVCYTKTWCDAFCSIRGKRVDLGCAATCPTVK −TGVDIQCCST −DNCNPFPT −RKRP
33. N. n. n., toxin 3	IRCF−−−ITPDITSKDCPNG −HVCYTKTWCDGFCSIRGKRVDLGCAATCPTVK −TGVDIQCCST −DNCNPFPT −RKRP
34. N. n. n., toxin 3	IRCF−−−ITPDITSKDCPNG −HVCYTKTWCDGFCSIRGKRVDLGCAATCPTVR −TVGDIQCCST −DNCNPFPT −RKRP
35. N. n. n., toxin 4	IRCF−−−ITPDITSKDCPNG −HVCYTKTWCDGFCSSRGKRVDLGCAATCPTVR −TGVDIQCCST −DNCNPFPT −RKRP
36. N. n., toxin-A	IRCF−−−ITPDITSKDCPNG −HVCYTKTWCDGFCSIRGKRVDLGCAATCPTVR −TGVDIQCCST −DDCDPFPT −RKRP
37. N. n., toxin-B	IRCF−−−ITPDITSKDCPNG −HVCYTKTWCDGFCSSRGKRVDLGCAATCPTVR −TGVDIQCCST −DDCDPFPT −RKRP
38. N. n., toxin-C	IRCF−−−ITPDITSKDCPNG −HVCYTKTWCDAFCSIRGKRVDLGCAATCPTVK −TGVDIQCCST −DDCDPFPT −RKRP
39. N. melan., toxin b	IRCF−−−ITPDVTSQICADG −HVCYTKTWCDNFCASRGKRVDLGCAATCPTVK −PGVNIKCCST −DNCNPFPT −RNRP
40. N. niv., toxin α	IRCF−−−ITPDVTSQACPDG −HVCYTKMWCDNFCGMRGKRVDLGCAATCPKVK −PGVNIKCCSR −DNCNPFPT −RKRS
41. N. h. h., CM-5	IRCF−−−ITPDVTSQACPDG −HVCYTKMWCDNFCGMRGKRVDLGCAATCPTVK −PGVDIKCCST −DNCNPFPT −RKRS
42. N. haj., toxin III	IRCF−−−ITPDVTSQACPDGQNICYTKTWCDNFCGMRGKRVDLGCAATCPTVK −PGVDIKCCST −DNCNPFPT −RERS
43. D. j. kai., $V_N^{III}1$	RTC YY−−KTYSDKSKTCPRGENICYTKTWCDGFCSQRGKRVELGCAATCPKVK −TGVEIKCCST −DYCNPFPVW −NPR
44. D. p. poly., toxin γ	RTC−N−−KTFSDQSKICPPGENICYTKTWCDAWCSQRGKRVELGCAATCPKVK −AGVEIKCCST −DDCDKFQF −GKPR
45. D. p. poly, toxin δ	RTC−N−−KTPSDQSKICPPGENICYTKTWCDAWCSQRGKIVELGCAATCPKVK −AGVEIKCCST −DNCNKFKF −GKPR
46. D. p. poly., V_N2	RTC−N−−KTFSDQSKICPPGENICYTKTWCDAWCSRRGKIVELGCAATCPKVK −AGVGIKCCST −DNCNLFKF −GKPR
47. D. vir., toxin I	RTCY−−−KTPSVKPETCPHGENICYTETWCDAWCSQRGKRVELGCAATCPKVK −AGVGIKCCST −DNCNPFPVW −NPRG
48. D. vir., toxin V	RTCY−−−KTPSVKPETCPHGENICYTETWCDAWCSQRGKRVELGCAATCPKVK −AGVGIKCCST −DNCNPFPVW −NPR
49. D. vir., toxin 4.7.3. 4.9.3.	RTCY−−−KTPSVKPETCPHGENICYTETWCDAWCSQRGKREELGCAATCPKVK −AGVGIKCCST −DNCDPFPV −KNPR
50. N. n. ox., neurotox. I	ITCY −−−KTPIPITSETCAPGQNLCYTKTWCDAWCGSRGKVIELGCAATCPTVE −SYQDIKCCST −DDCNPHPK −QKRP
51. O. hann., toxin a	TKCY −−−VTPDVKSETCPAGQDICYTETWCDAWCTSRGKRVDLGCAATCPIVK −PGVEIKCCST −DNCNPFPTWRKRP
52. O. hann., toxin b	TKCY −−−VTPDATSQTCPDGENICYTKTWCDGFCSSRGKRIDIGCAATCPKVK −PGVDIKCCST −DNCNPFPTWKRKH
53. B. m., α-bungarotoxin	IVCH −TTATIPSSAVTCPPGENLCYRKMWCDAFCSSRGKVVEBGCAATCPSKK −PYEEVTCCST −DKCNHPRK −RQPG
54. N. mela., toxin 3.9.4.	KRCY −−−RTPNLKSQTCPPGEDLCYTKKWCDAWCTSRGKVIELGCVATCPKVK −PYEQITCCST −DNCNPHPK −MKP
55. L. sem. Ls III	RECY −−−LNPHDT−QTCPSGQEICYVKSWCNAWCSSRGKVLEFGCAATCPSVN −TGTEIKCCSA −DKCNTYP
56. Not. s. s., III-4	LICY −−−−MGPKTPRTCPRGQNLCYTKTWCDAFCSSRGKVVELGCAATCPIAK −SYEDVTCCST −DNCNPFPV −RPRHPP
57. N. h. ann., CM-2a	LECY−−−−−QMSKVVTCKPEETFCYSDVFMP −−−FRNHIVYTSGCSSYCRDGT −GEK−−−CCTT −DRCNGARGG
58. N. h., ann., CM-3	LECY−−−−−QMSKVVTCKPEEKFCYSDVFMP −−−FRNH−VYTSGCSSYCRDGT −GEK−−−CCTT −DRCNGARGG
59. D. ang., toxin F 7	TMCYSHTTTSRAILTNC −−GENSCYRKSRRHP −−−−PKMVLGRGC −−GCPPGD −DNLEVKCCTSPDKCNY
60. D. p. poly., toxin C	TICYSHTTTSRAILKDC −−GENSCYRKSRRHP −−−−PKMVLGRGC −−GCPPGD −DYLEVKCCTSPDKCNY
61. D. ang., toxin F 8	MICYSHKTPQPSATITCE −EKT−CYKKSV−RKL −−−PAVVAGRGC −−GCPSKE −MLVAIHCCRS −DKCNE
62. D. vir., toxin 4.9.6.	MICYSHKTPQNSATITCE −EKT−CYK−FV−TKL −−−PGVILARGC −−GCPKKEIFRKSIHCCRS −DKCNE
63. D. j. kai., S_5C_4	MICYSHKTPQNSATITCE −EKT−CYKKFV−TNV −−−PGVILARGC −−GCPKKEIFR−SIHCCRS −DKCNE
64. N. melan., S_4C_{11}	LTCLICPEKYCNKVHTCRNGENICFKRFYEGNL −−−LGKRYPRGCAATCPEAK −PREIVECCST −DKCNH
65. N. h. annn., CM-13b	LTCFNCPEVYCNRFHTCRNGEKICFKRFNERKL −−−LGKRYPTGCAATCPVAK −PREIVECCST −DRCNH
66. N. h. h., CM-11	LTCFICPEKYCNKVHTCRNGENQCFKRFNERKL −−−LGKRYTRGCAATCPEAK −PREIVECCTT −DRCNK

67. O. han., DE-1	LICFNQETYRPETTTTCPDGEN–CYSTFWHN –––DGHVKIERGC––GCPRVNPPISI––CCKT–DKCNN
68. D. p. poly., FS 2	RICYSHKASLPRATKTCV––ENTCYKMFIRT –––HRQYISERGC––GCPTAMWPYQT––ECCKG––DRCNK
69. D. j. kai., S_2C_4	LTCVTDKSFGGVNTEECAAGQKICFKNWKKMGPK ––LYDV–KRGCTATCPKADDDG–CVKCCNT–KG
70. N. melan., $V^{II}1$	LEC–N –KLVPIAHKTCPAGKNLCY–QMYMVS –––KSTIPVKRGCIDVCPKSS –LLVKYVCCNT –DRCN
70a. N. melan., $V^{II}1A$	
71. N. h. ann., CM-2e	LEC–N– –KLVPIAHKTCPEGKNLCY –KMFMVS –––TSTVPVKRGCIDVCPKDS –ALVKYVCCNT–DRCN
72. N. h. ann., CM-4a	LEC–N– –QLIPIAHKTCPEGKNLCY –KMFMVS –––TSTVPVKRGCIDVCPKNS –ALVKYVCCNT–DRCN
73. N. n. ox., cytotoxin	LKC–K– –KLVPLFSKTCPAGKNLCY –KMFMVA –––APHVPVKRGCIDVCPKSS –LLVKYVCCNT –DKCN
74. N. niv., $V^{II}3$	LKC–N– –QLIPPFWKTCPKGKNLCY –NMYMVS –––TSTVPVKRGCIDVCPKNS –ALVKYVCCNT–DRCN
75. N. nigr., toxin F-14	LKC–N– –QLIPPFWKTCPKGKNLCY –KMTMRA –––APMVPVKRGCIDVCPKSS –LLIKYMCCNT –DKCN
75a. N. nigr., cardiotoxin	
76. N. m. moss., $V^{II}1$	LKC–N– –QLIPPFWKTCPKGKNLCY –KMTMRA –––APMVPVKRGCIDVCPKSS –LLIKYMCCNT –NKCN
77. N. m. moss., $V^{II}2$	LKC–N– –QLIPPFWKTCPKGKNLCY –KMTMRG –––ASKVPVKRGCIDVCPKSS –LLIKYMCCNT –DKCN
78. N. m. moss., $V^{II}3$	LKC–N– –RLIPPFWKTCPEGKNLCY –KMTMRL –––APKVPVKRGCIDVCPKSS –LLIKYMCCNI –NKCN
79. N. h. ann., CM-11	LKC–N– –KLIPPFWKTCPKGKNLCY –KMYMVS –––TLTVPVKRGCIDVCPKNS –ALVKYVCCNT–NKCN
80. N. n. atra, cardiotoxin	LKC–N– –KLVPLFYKTCPAGKNLCY–KMFMVA –––TPKVPVKRGCIDVCPKSS –LLVKYVCCNT –DRCN
81. N. n. a., cardiotoxin II	LKC–N– –KLVPLFYKTCPAGKNLCY–KMFMVS –––NLTVPVKRGCIDVCPKNS –ALVKYVCCNT–DRCN
82. N. n. a., cardiotoxin IV	RKC–N– –KLVPLFYKTCPAGKNLCY–KMFMVS –––NLTVPVKRGCIDVCPKNS –ALVKYVCCNT–DRCN
83. N. n. n., cytotoxin II	LKC–N– –KLVPLFYKTCPAGKNLCY–KMYMVA –––TPKVPVKRGCIDVCPKSS –LVLKYVCCNT –DRCN
84. N. n. n., cytotoxin IIa	LKC–N– –KLIPLAYKTCPAGKNLCY –KMFMVS –––NKTVPVKRGCIDVCPKNS–LVLKYVCCNT –DRCN
85. N. naja, CM-XI	LKC–N– –KLIPLAYKTCPAGKNLCY –KMYMVS –––TPKVPVKRGCIDVCPKNS –LVLKYECCNT –DRCN
86. N. naja, cytotoxin I	LKC–N– –KLIPLAYKTCPAGKNLCY –KMYMVS –––NKTVPVKRGCIDVCPKNS–LVLKYECCNT –DRCN
87. N. m. moss., $V^{II}4$	LKC–N– –KLIPIAYKTCPEGKNLCY –KMMLAS –––KKMVPVKRGCINVCPKNS–ALVKYVCCST –DRCN
88. N. n. a., cardiotoxin I	LKC–N– –KLIPIASKTCPAGKNLCY –KMFMMS –––DLTIPVKRGCIDVCPKSN –LLVKYVCCNT–DRCN
88a. N. n., cardiotoxin	
89. N. n., toxin F-8	LKC–N– –KLIPIASKTCPAGKNLCY –KMFMMS –––DLTIPVKRGCIDVCPKNS –LLVKYVCCNT–DRCN
90. N. h. ann., CM-2h	LKCH–– –KLVPPFWKTCPEGKNLCY–KMYMVA –––TPMLPVKRGCIDVCPKDS –ALVKYMCCNT–NKCN
91. N. h. ann., CM-2hA	LKCH–– –KLVPPFWKTCPEGKNLCY–KMYMVA –––TPMLPVKRGCIDVCPKDS –ALVKYVCCST –NKCN
92. N. h. ann., CM-4b	LKCH–– –KLVPPFWKTCPEGKNLCY–KMYMVA –––TPMLPVKRGCINVCPKDS –ALVKYMCCNT–NKCN
93. N. h. ann., CM-4bA	LKCH–– –KLVPPFWKTCPEGKNLCY–KMYMVA –––TPMLPVKRGCINVCPKDS –ALVKYVCCST –NKCN
94. N. h. ann., CM-6	LKCH–– –KLVPPFWKTCPEGKNLCY–KMYMVY –––TPMIPVKRGCIDVCPKNS –ALVKYMCCNT–NKCN
95. N. h. ann., CM-7	LKCH–– –KLVPPFWKTCPEGKNLCY–KMYMVS –––TLTVPVKRGCIDVCPKNS –ALVKYVCCNT–NKCN
96. N. hann., $V^{II}1$	LKCH–– –KLVPPVWKTCPEGKNLCY––KMFMVS–––TSTVPVKRGCIDVCPKNS –ALVKYVCCST –DKCN
97. N. h. ann., $V^{II}2$	LKCH–– –KLVPPFWKTCPEGKNLCY–KMYMVA –––TPMLPVKRGCIDVCPKDS –ALVKYMCCNT–DKCN
98. N. h. ann., $V^{II}2A$	LKCH–– –KLVPPFWKTCPEGKNLCY–KMYMVA –––TPMLPVKRGCIDVCPKDS –ALVKYVCCST –DKCN
99. N. h. ann., CM-8	LKCY–– –KLVPPFWKTCPEGKNLCY–KMYMVS –––TLTVPVKRGCIDVCPKNS –ALVKYVCCNT–DKCN

100. N. h. ann., CM-8A	LKCH− − −KLVPPFWKTCPEGKNLCY−KMYMVS − − −TLTVPVKRGCIDVCPKNS −ALVKYVCCNT−DKCN
101. N. niv., $V^{II}1$	LKCH− − −KLVPPVWKTCPEGKNLCY−KMFMVS − − −TSTVPVKRGCIDVCPKDS −ALVKYVCCST −DKCN
102. N. niv., $V^{II}2$	LKCH− − −QLIPPFWKTCPEGKNLCY −KMYMVA − − −TPMIPVKRGCIDVCPKNS −ALVKYMCCNT−DKCN
103. N. h. h., CM-7	LKCH− − −QLVPPFWKTCPEGKNLCY−KMYMVA − − −TPMIPVKRGCIDVCPKNS −ALVKYMCCNT−DKCN
104. N. h. h., CM-8	LKCH− − −QLVPPFWKTCPEGKNLCY−KMYMVS − − −SSTVPVKRGCIDVCPKNS −ALVKYCCNT −DKCN
105. N. h. h., CM-9	LKCH− − −QLVPPFWKTCPAGKNLCY−KMYMVA − − −TPMIPVKRGCIDVCPKNS −ALVKYMCCNT−DKCN
106. N. h. h., CM-10b	LKCH− − −KLVPPFWKTCPEGKNLCY−KMYMVA − − −TPMIPVKRGCIDVCPKNS −ALVKYVCCNT−DKCN
107. H. hem., DLF (F 12B)	LKCHN− −KLVPFLSKTCPEGKNLCY −KMTMLK − − −MPKIPIKRGCTDACPKSS −LLVKVVCCNK−DKCN
108. N. h. ann., CM-13a	LKCHN− −TQLPFIYKTCPEGKNLCF −KTTLKKL − −PLKIPIKRGCAATCPKSS −ALLKVVCCST −DKCN
108a. N. h. h., CM-12	
109. H. haem., toxin 11	LKCHN− −KLVPFLSKTCPDGKNLCY −KMSMEV − − −TPMIPIKRGCTDTCPKSS −LLVKVVCCKT −DKCN
110. H. haem., toxin 11A	LKCHN− −KLVPYLSKTCPDGKNLCY−KMSMEV − − −TPMIPIKRGCTDTCPKSS −LLVKVVCCKT −DKCN
111. H. haem., toxin 12A	LKCHN− −KVVPFLSKTCPEGKNLCY −KMTLKK − − −VPKIPIKRGCTDACPKSS −LLVNVMCCKT−DKCN
112. H. haem., toxin 9B	LICHN− −RPLPFLHKTCPEGQNICY −KMTLKK − −TPMKLSVKRGCAATCPSAR −PLVQVECCKT −DKCNW
113. H. haem., toxin 9BB	LICHN− −RPLFLHKTCPEGQNICY −KMTLKK − −TPMKLSVRRGCAATCPSAR −PLVQVECCKT −DKCNW
114. N. melan., $V^{II}2$	IKCHN− −TLLPFIYKTCPEGQNLCF −KGTLKF − − −PKKTTYNRGCAATCPKSS −LLVKYVCCNT −NKCN
115. N. melan., $V^{II}3$	IKCHN− −TLLPFIYKTCPEGQNLCF −KGTLKF − − −PKKTTYKRGCAATCPKSS −LLVKYVCCNT −NKCN
116. N. melan., toxin 3.20	IKCHN− −TPLPFIYKTCPEGNNLCF −KGTLKF − − −PKKITYKRGCADACPKTS −ALVKYVCCNT−DKCN
117. Porcine pancreas	ALWQFRSMINCAIPGSHPLMDFNNYGCYCGLGGSGTPVDELDRCCETHDNCYRDAKNLDSCKFLVDNPYT
118. β_1-Bungarot., A-chain	NLINFMEMIRYTIPCEKTWGEYADYGCYCGAGGSGRPIDALDRCCYVHDNCYGDAEKKHKCNPKTSQ−YS
119. Notexin	NLVQFSYLIQCANHGKRPTWHYMDYGCYCGAGGSGTPVDELDRCCKIHDDCYDEAGKK−GCFPKMSA− − −
120. Notechis 5	NLVQFSYLIQCANHGRRPTRHYMDYGCYCGWGGSGTPVDELDRCCKIHDDCYSDAEKK −GCSPKMSA− − −
121. Bitis gabonica	DLTQFGNMIN− − −KMGQSVFDYIYYGCYCGWGGKGKPIDATDRCCFVHDCCYGKMGTYDT−KWTSYN− − −
122. N. melanoleuca, DE-III	NLYQFKNMIHCTVPNR−SWWHFANYGCYCGRGGSGTPVDDLDRCCQIHDNCYGEAEKISGCWPYIKT−YT
123. H. haemachatus, DE-I	NLYQFKNMIKCTVPSR−SWWHFANYGCYCGRGGSGTPVDDLDRCCQTHDNCYSDAEKISGCRPYFKT−YS
	ESYSYSCSNTEITCNSKNNA−CEAFICNCDRNAAICFSKAPYNKEHKNLDTKKY−C
	− −YKL− −TKRTIICYGAAGGTC−RIVCDCDRTAALCFGQSDYIEEHKNIDTARF−CQ
	− −YDYYCGENGPYCRNIKKK −CLRFVCDCDVEAAFCFAKAPYNNANWNIDTKKR−CQ
	− −YDYYCCENGPYCRNIKKK −CLRFVCDCDVEAAFCFAKAPYNNANWNIDTKKR−CQ
	− −YEI− −QNGGIDCDEDPQK − −KEL−CECDRVAAICFANNRNTYNSNYFGHSSSKCTGTEQC
	− −YD−SCQGTLTSCGAANN− −CAASVCDCDRVAANCFARAPYIDKNYNIDFNAR−CQ
	− −YD− −CTKGKLTCKEGNNE −CAAFVCKCDRLAAICFAGAHYNDNNNYIDLARH−CQ
124. Bovine pancreatic inh.	RPDFCLEPPYTGPCKARIIRYFYNAKAGLCQTFVYGGCRAKRNNFKSAEDCMRTCGGA
125. Vipera ruselli inhibitor	HDRPTFCNLAPESGRCRGHLRRIYYNLESNKCKVFFYGGCGGNANNFETRDECRETCGGK
126. Hemach. haem. inhibitor	RPDFCELPAETGLCKAYIRSFHYNLAAQQCLQFIYGGCGGNANRFKTIDECRRTCVG

127. Naja nivea inhibitor	RPRFCELPAETGLCKARIRSFHYNRAAQQCLEFIYGGCGGNANRFKTIDECHRTCVG
128. Dendro. p. poly., toxin K	AAKYCKLPLRIGPCKRKIPSFYYKWKAKQCLPFDYSGCGGNANRFKTIEECRRTCVG
129. Dendro. p. poly,., toxin E	LQHRTFCKLPAEPGPCKASIPAFYYNWAAKKCQLFHYGGCKGNANRFSTIEKCRHACVG
130. β_1-Bungarotoxin, B-chain	RQRHRDCDKPPDKGNCGPV−RAFYYDTRLKTCKAFQYRGCDGDHGNFKTETLCR−−CECLVYP
131. Cr. d. t., crotamine	YKQCHKKGGHCFPKEKICLPPSSDFGKMDCRWRWKCCKKGSG
132. Cr. v. h., peptide C	YKRCHKKGGHCFPKTVICLPPSSDFGKMDCRWKWKCCKKSVN
133. Cr. v. v., myotoxin a	YKQCHKKGGHCFPKEKICIPPSSDLGKMDCRWKWKCCKKGSG

After completion of the list the following sequence data have been received:

Neurotoxins:	
134. Astr. stok., toxin a	MTCCNQQSSQPKTTTNC−A−GNSCYKKTWSD−H−−−RGTIIERGC−−GCPQVK −SGIKLECCHT −NECNN
135. Astr. stok., toxin b	LSCY−−−LGY−KHSQTCPPGENVCFVKTWCDGFCNTRGERIIMGCAATCPTAK −SGVHIACCST −DNCNIYAKWGS(NH_2)
136. Astr. stok., toxin c	LSCY−−−LGY−KHSQTCPPGENVCFVKTWCDAFCSTRGERIVMGCAATCPTAK −SGVHIACCST −DNCNIYTKWGSGR(NH_2)
Phospholipase A	
137. B. fasciatus, cardiotoxin	NLYQFKNMIECAGTRNIAGFTNWQALVKKYGCYCGPGTHDPDALEKNGCYTGPLRFGNIY NLAAKCCGSPNRKTYVYTCNAPAFGIKTVCDCDRDCQTCDAYHKTALATGIDETKHCQ

7.1 Elapidae

The elapid snakes are the most common family of venomous snakes, they are distributed in America, Asia, Africa and Australia.

In *America* the elapids are represented by the coral snakes of the genera *Micruroides* and *Micrurus*. In North America *Micruroides euryxanthes* (Western Coral Snake) and *Micrurus fulvius* (Eastern Coral Snake) are especially worthy of mention.

Outside the USA, the Eastern Coral Snake can be found in Mexico, Central America and South America down to the northern parts of Argentina. In South America three other important species have to be added: *M. corallinus, M. frontalis* (both in Brazil) and *M. lemniscatus* (Northern South America, Trinidad and Amazon region). The bright and colorful appearance of alternating red, yellow, white and black rings is most striking. The animals are small: *Micruroides,* 40–50 cm; *Micrurus* 60–90 cm. They are likely to bite, even if there is no threat.

In *Africa* three elapids should be mentioned because of their wide distribution (except North Africa) and the danger they pose.

1. *Naja nigricollis.* This species is commonly known as the spitting cobra. Cobras usually live in a given place: between rocks, in holes or abandoned termite burrows. They rest during the day and hunt after dark. They may enter houses searching for small mammals. The spitting cobra does not like to bite; rather, it raises its head and spits the venom at the victim. This snake is rather unerring over a distance of two meters.
2. *Haemachatus haemachatus.* This species is known as Ringhals cobra; it is very similar in habits to *Naja nigricollis.*
3. *Dendroaspis polylepis.* The "Black Mamba" has habits similar to those mentioned above. It is occasionally found in trees. If frightened by a human, it tries to drive him away by raising its body and opening its mouth making the black throat visible. Any movement will now cause the snake to strike. The bite usually hits the head or the upper part of the body and in most cases is fatal. In an encounter with this snake one should remain motionless until it lowers its body and goes away. The black mamba is nocturnal, and accidents, fortunately, are rare.

Among the elapids of *Asia,* two genera are abundant and widely distributed: *Naja* (cobra) and *Bungarus* (krait). Within these genera the following species are principally responsible for accidents: *N. naja, B. fasciatus* and *B. caeruleus.* The toxins of the Asian and African cobras differ in that the Asian species possess an additional cardiotoxin. The toxins of kraits are neurotoxic.

B. multicinctus, N. naja and *Calliophis spp.* are found on the East Asian islands, Taiwan, Malaysia, the Philippines, as well as South China.

The dangerous snakes of Australia and Melanesia belong exclusively to the elapids. The most important species are:

Acantophis antarcticus (Death Adder)
Notechis scutatus (Tiger Snake)
Denisonia superba (Copperhead)
Oxyuranus scutellatus (Taipan)
Pseudechis porphyriacus (Red-bellied Black Snake)
Demansia textilis (Brown Snake)

Snakebites in Australia frequently are non-characteristic and occasionally without pain. However, one must assume that all bites come from venomous species. In any case a doctor should be seen as soon as possible.

Envenomotion

Micruroides species. Bites are rare, fatal cases unknown. Local symptoms consist of pain lasting up to six hours, as well as paresthesia in the afflicted limb. General symptoms include weakness, nausea and visual disturbances. All complaints disappear spontaneously within 24 hours.

Micrurus species. Envenomation from *Micrurus* is more severe; the death rate according to different statistics is 10–20%. Higher figures (up to 75%) occasionally published are certainly erroneous. The venom is an extremely potent neurotoxin, and in severe cases death may occur within four hours. All victims, therefore, should be hospitalized for at least two days. Bites are usually not painful; sometimes they go unnoticed, especially among children who are therefore particularly endangered. The local symptoms are, at

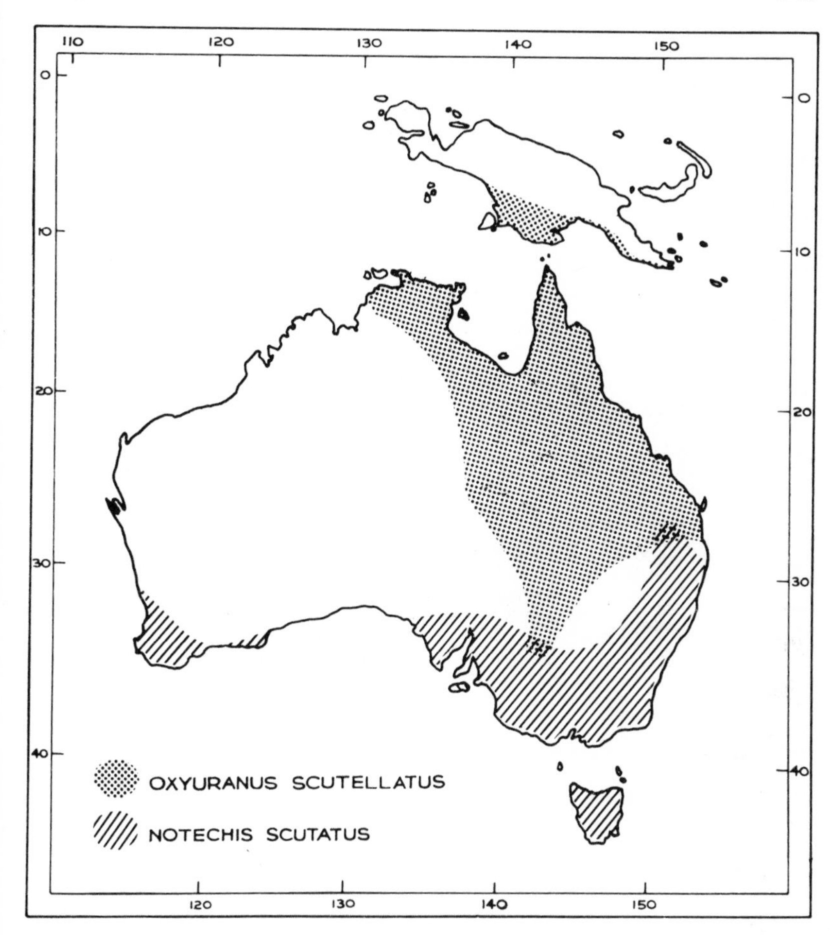

Fig. 20

Figs. 20–24. Distribution of venomous snakes in Australia (from H. G. Cogger)

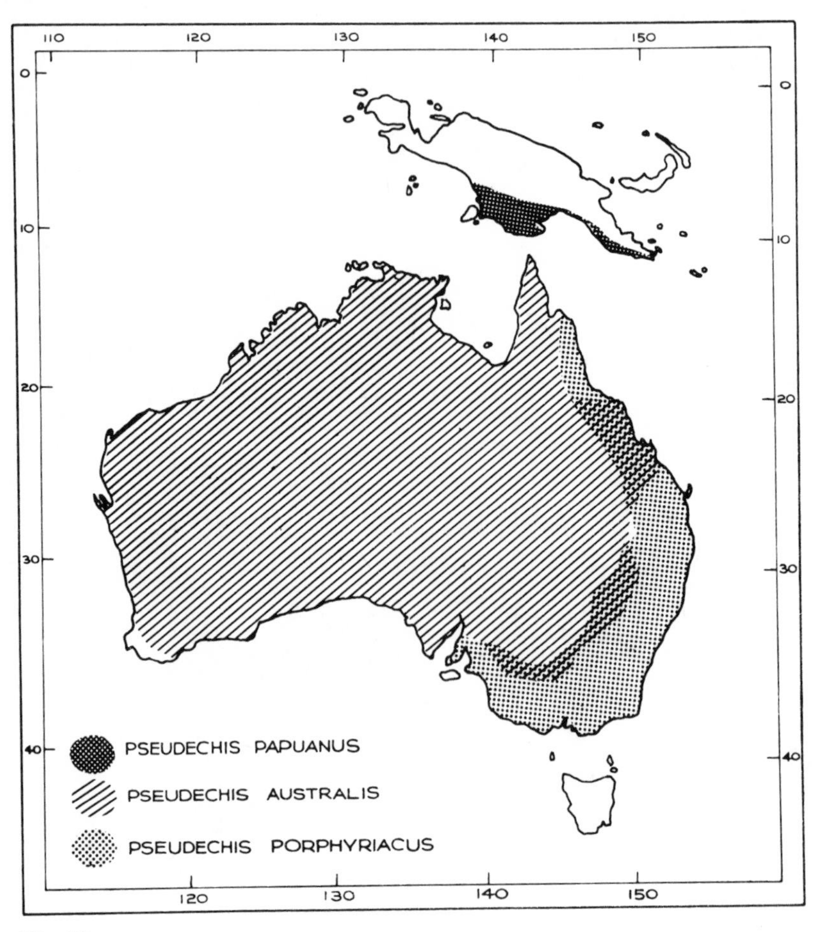
110
120
130
140
150
0
10
20
30
40
PSEUDECHIS PAPUANUS
PSEUDECHIS AUSTRALIS
PSEUDECHIS PORPHYRIACUS

Fig. 21

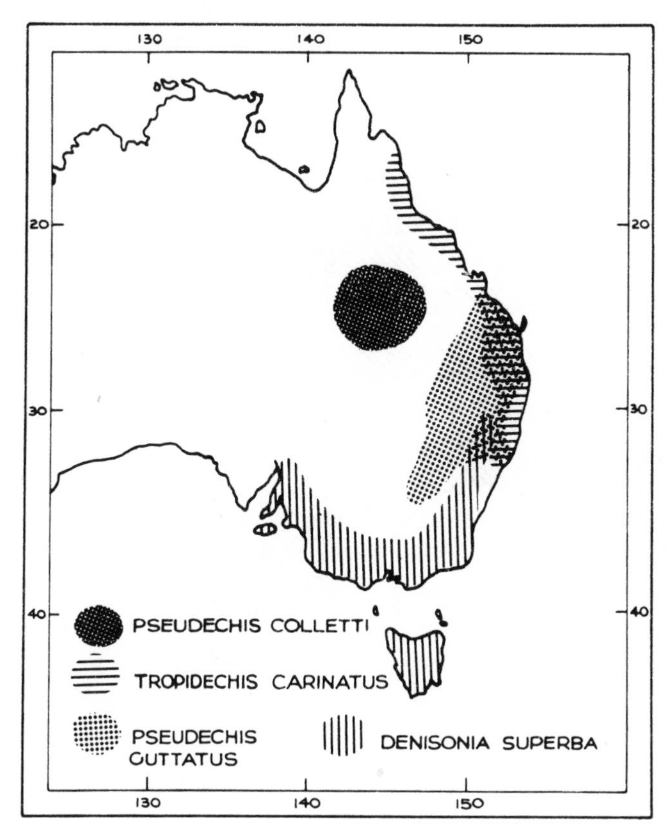
130
140
150
20
30
40
PSEUDECHIS COLLETTI
TROPIDECHIS CARINATUS
PSEUDECHIS GUTTATUS
DENISONIA SUPERBA

Fig. 22

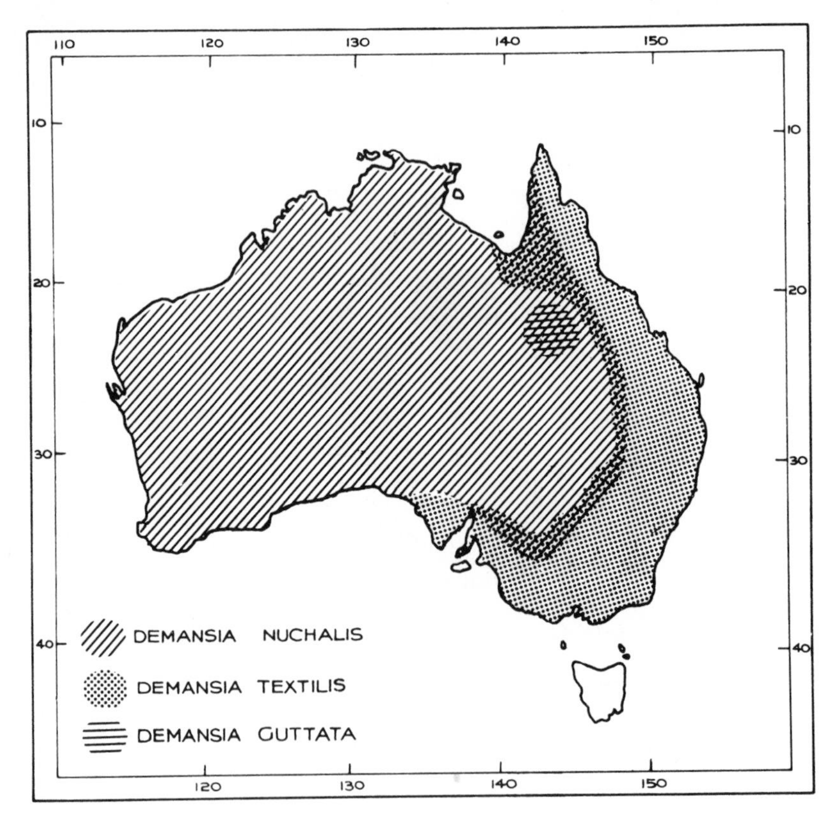
110
120
130
140
150
10
20
30
40
DEMANSIA NUCHALIS
DEMANSIA TEXTILIS
DEMANSIA GUTTATA

Fig. 23

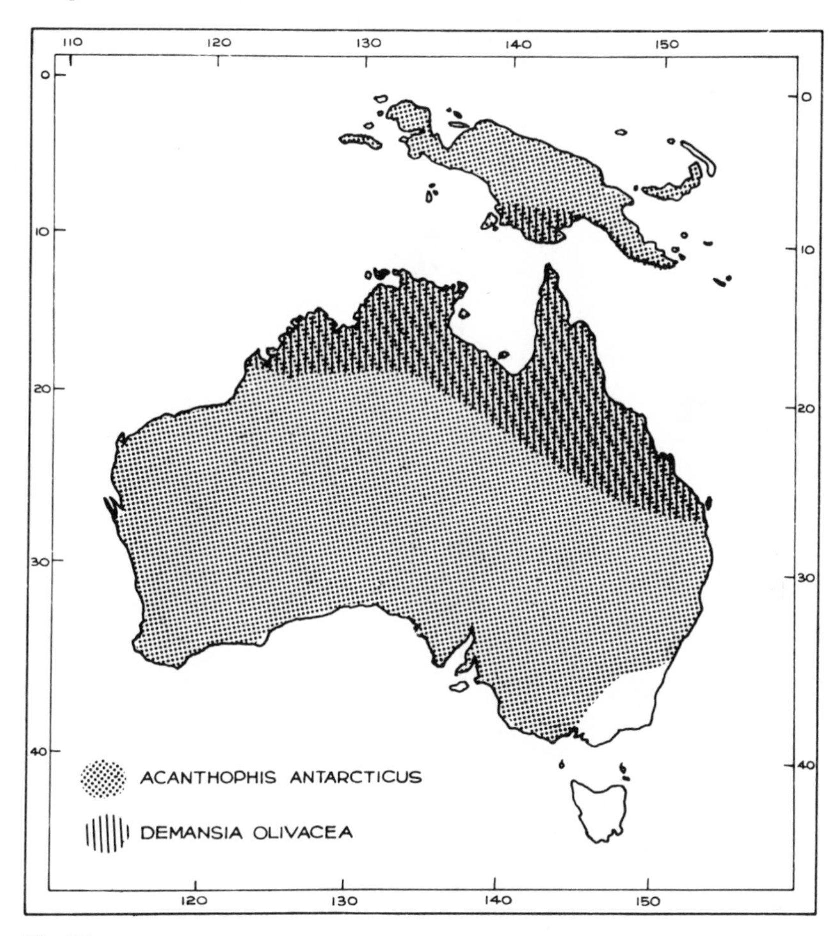
110
120
130
140
150
0
10
20
30
40
ACANTHOPHIS ANTARCTICUS
DEMANSIA OLIVACEA

Fig. 24

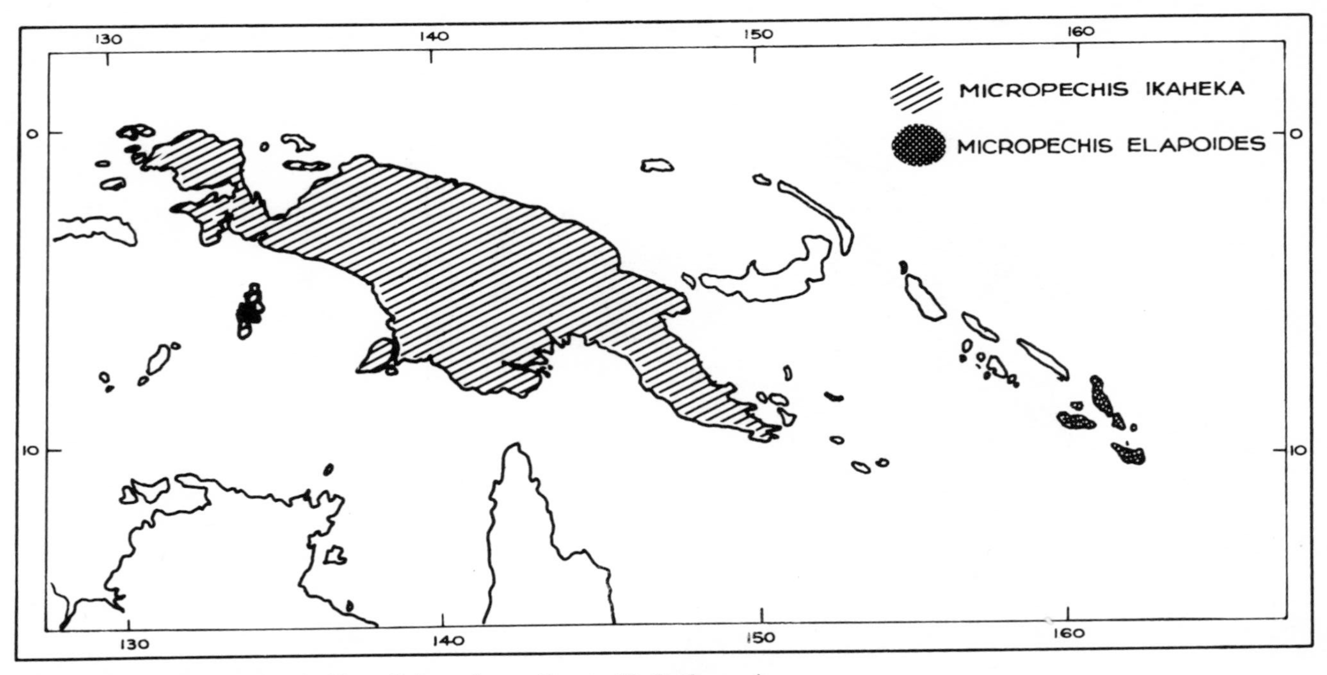

Fig. 25. Venomous snakes in New Guinea (according to H. G. Cogger)

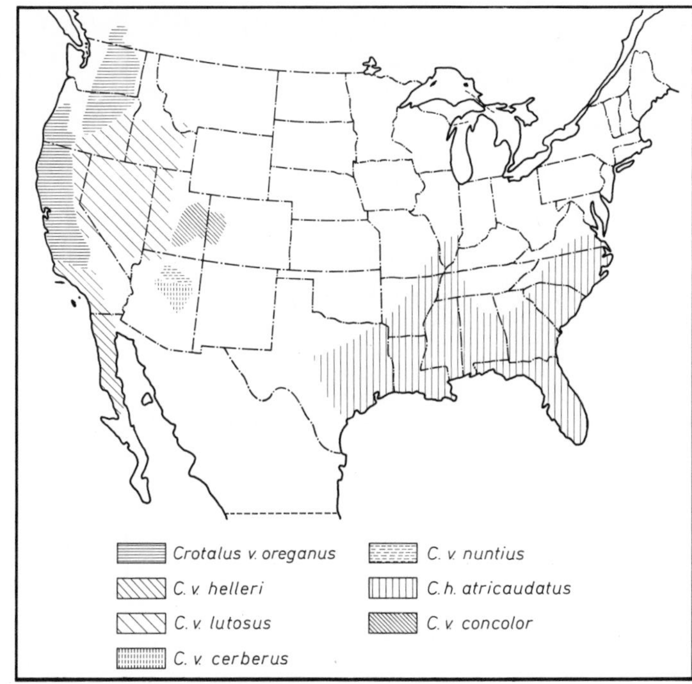
Crotalus v. oreganus
C. v. helleri
C. v. lutosus
C. v. cerberus
C. v. nuntius
C. h. atricaudatus
C. v. concolor

Fig. 26

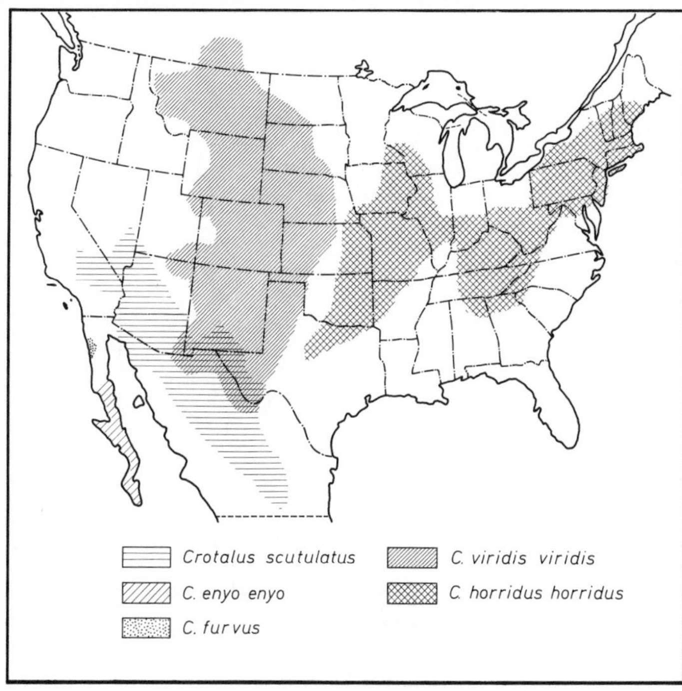
Crotalus scutulatus
C. enyo enyo
C. furvus
C. viridis viridis
C. horridus horridus

Fig. 27

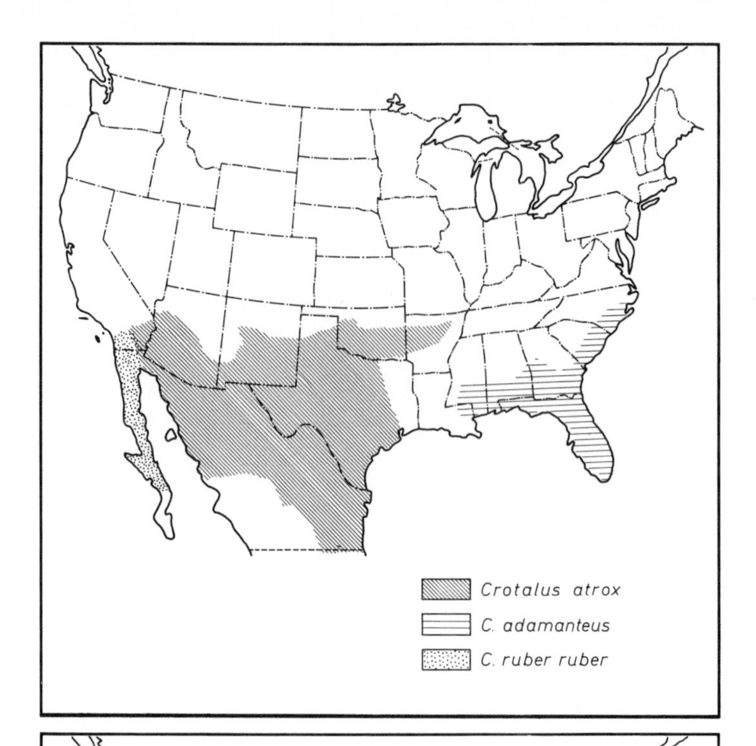
Crotalus atrox
C. adamanteus
C. ruber ruber

Fig. 28

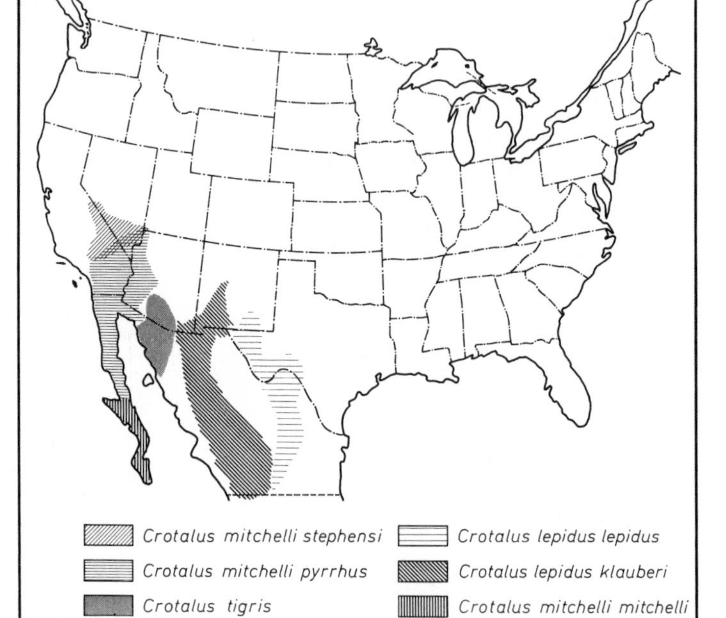
Crotalus mitchelli stephensi
Crotalus mitchelli pyrrhus
Crotalus tigris
Crotalus lepidus lepidus
Crotalus lepidus klauberi
Crotalus mitchelli mitchelli

Fig. 29

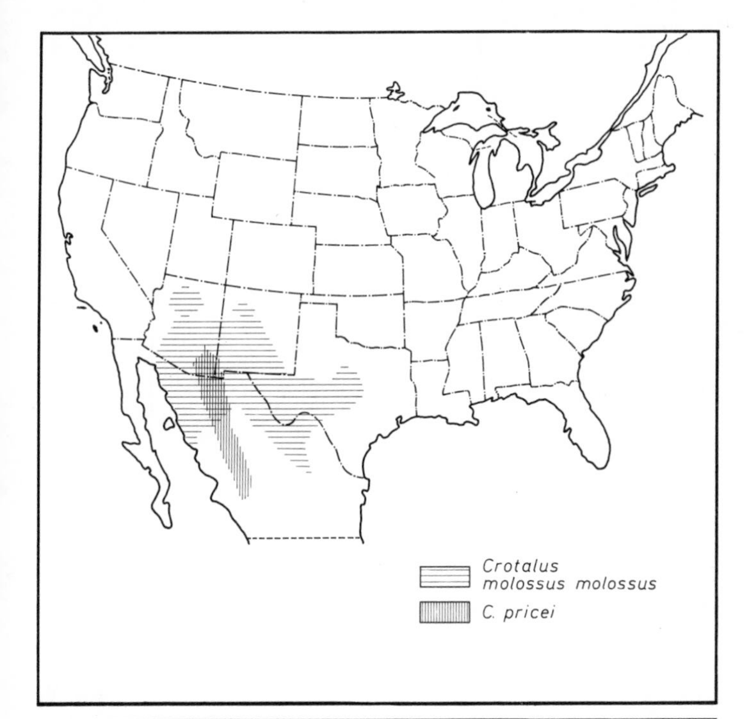
Crotalus
molossus molossus
C. pricei

Fig. 30

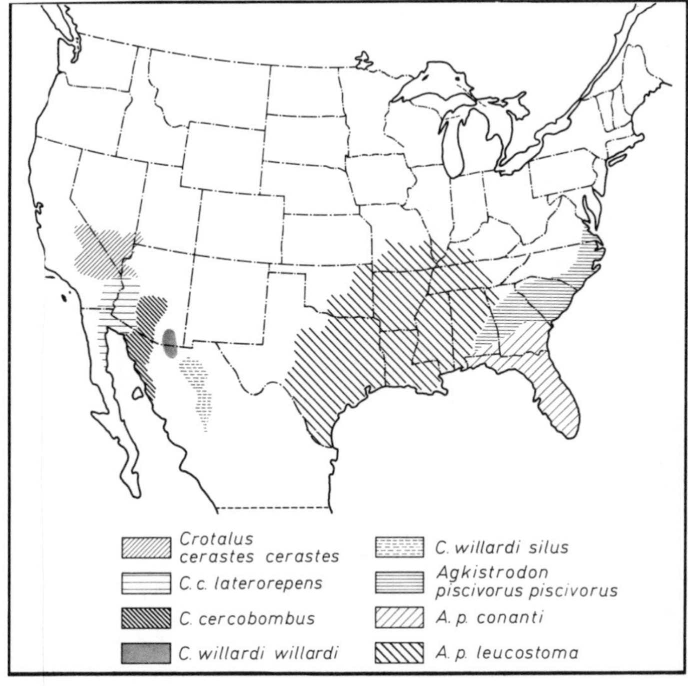
Crotalus
cerastes cerastes
C. c. laterorepens
C. cercobombus
C. willardi willardi
C. willardi silus
Agkistrodon
piscivorus piscivorus
A. p. conanti
A. p. leucostoma

Fig. 31

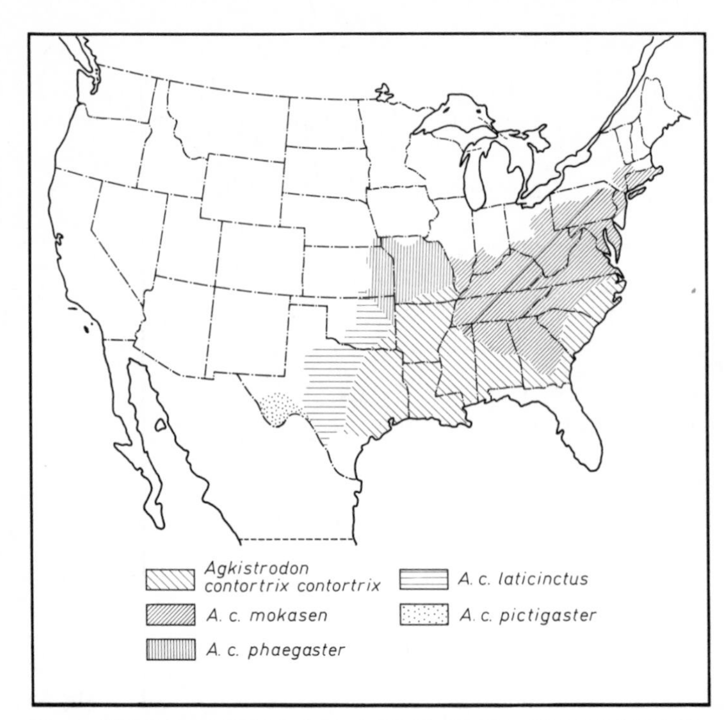

Agkistrodon contortrix contortrix
A. c. laticinctus
A. c. mokasen
A. c. pictigaster
A. c. phaegaster

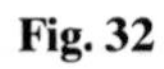
Fig. 32

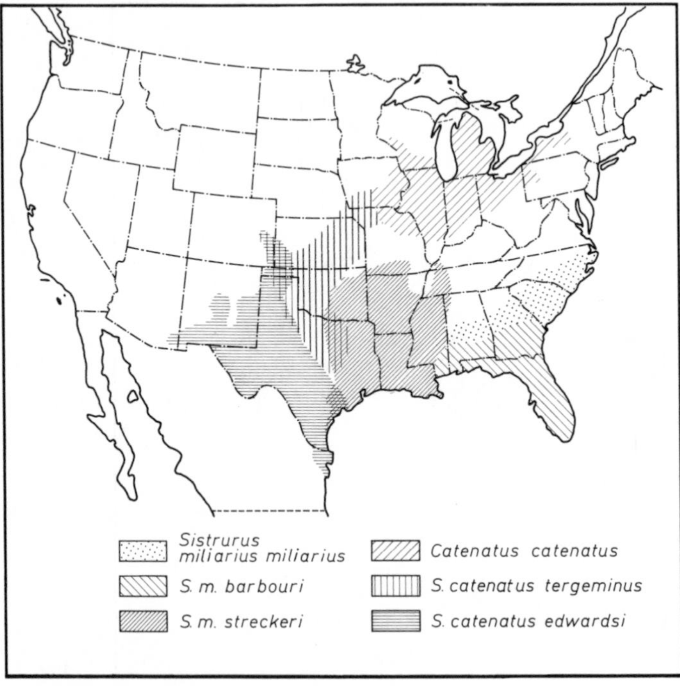

Sistrurus miliarius miliarius
Catenatus catenatus
S. m. barbouri
S. catenatus tergeminus
S. m. streckeri
S. catenatus edwardsi

Fig. 33

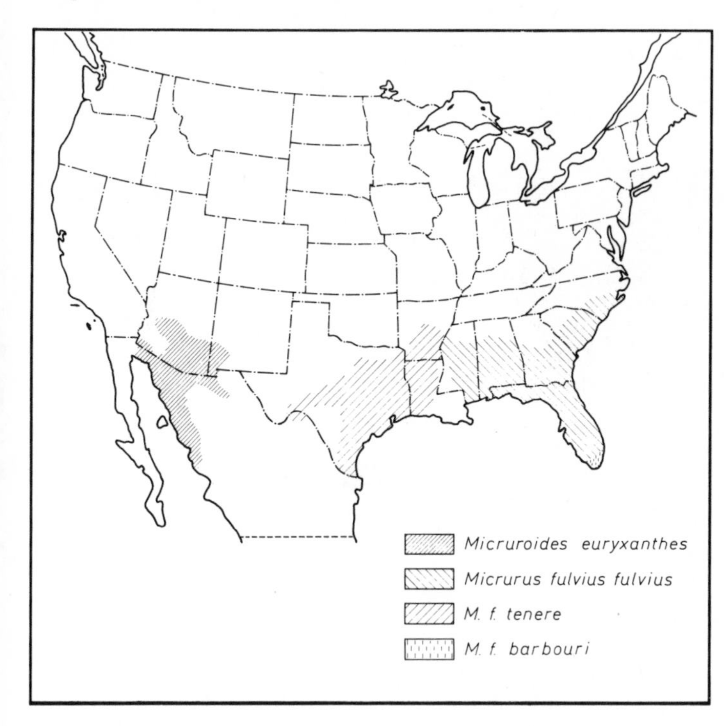

Fig. 34

Figs. 26–34. Distribution of venomous snakes in North America (according to F. E. Russell)

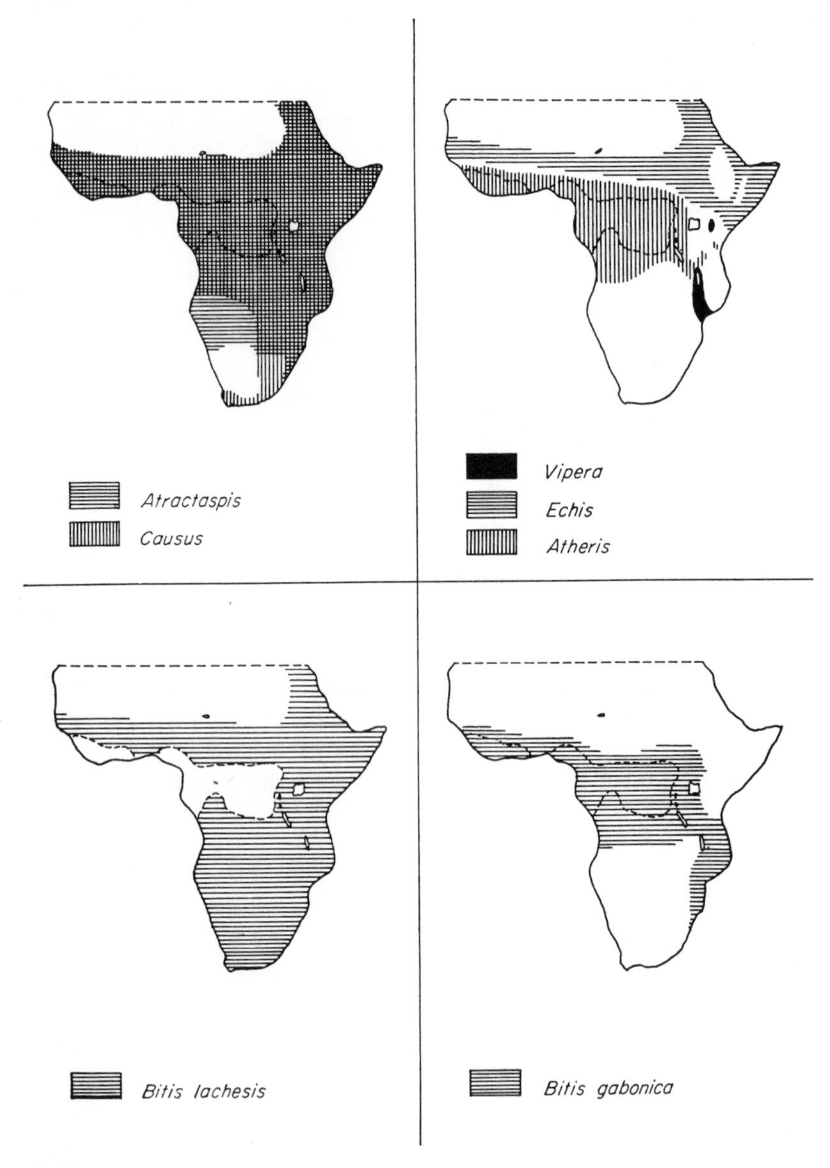

Fig. 35

Figs. 35–38. Distribution of venomous snakes in Central- and South Africa (according to D. C. Broadley)

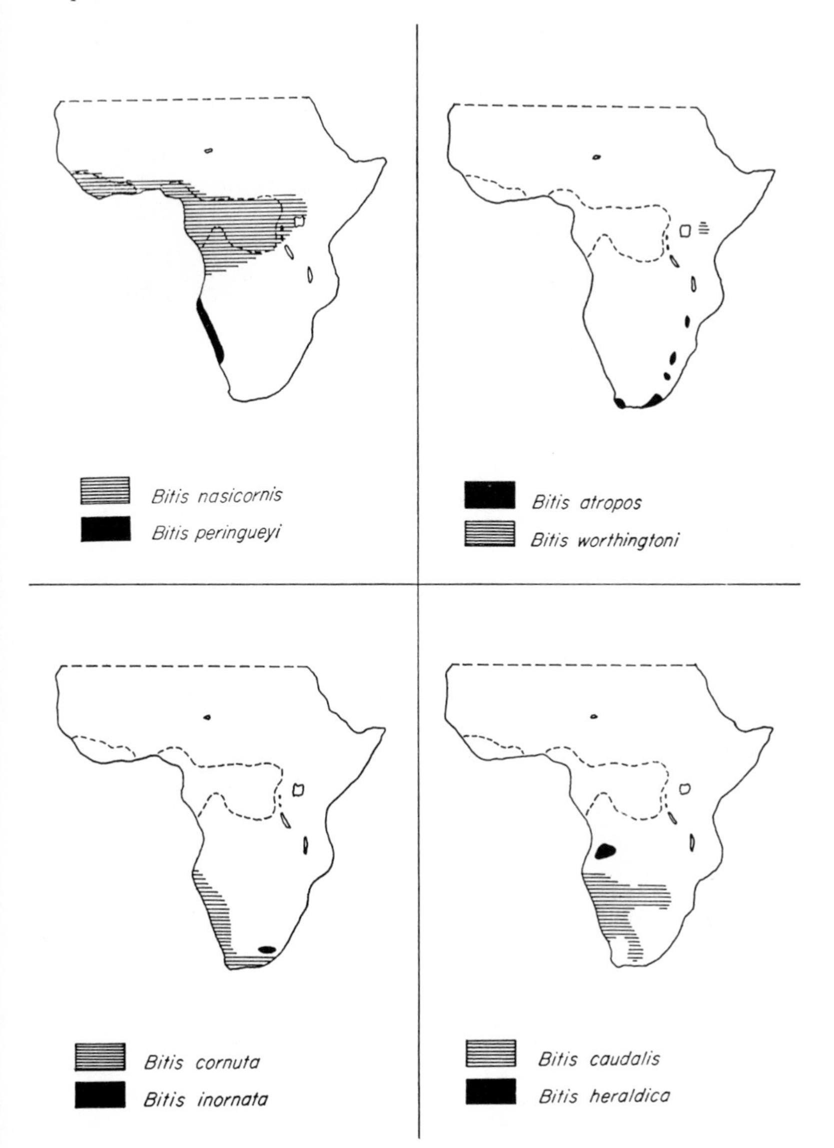
Bitis nasicornis
Bitis peringueyi
Bitis atropos
Bitis worthingtoni
Bitis cornuta
Bitis inornata
Bitis caudalis
Bitis heraldica

Fig. 36

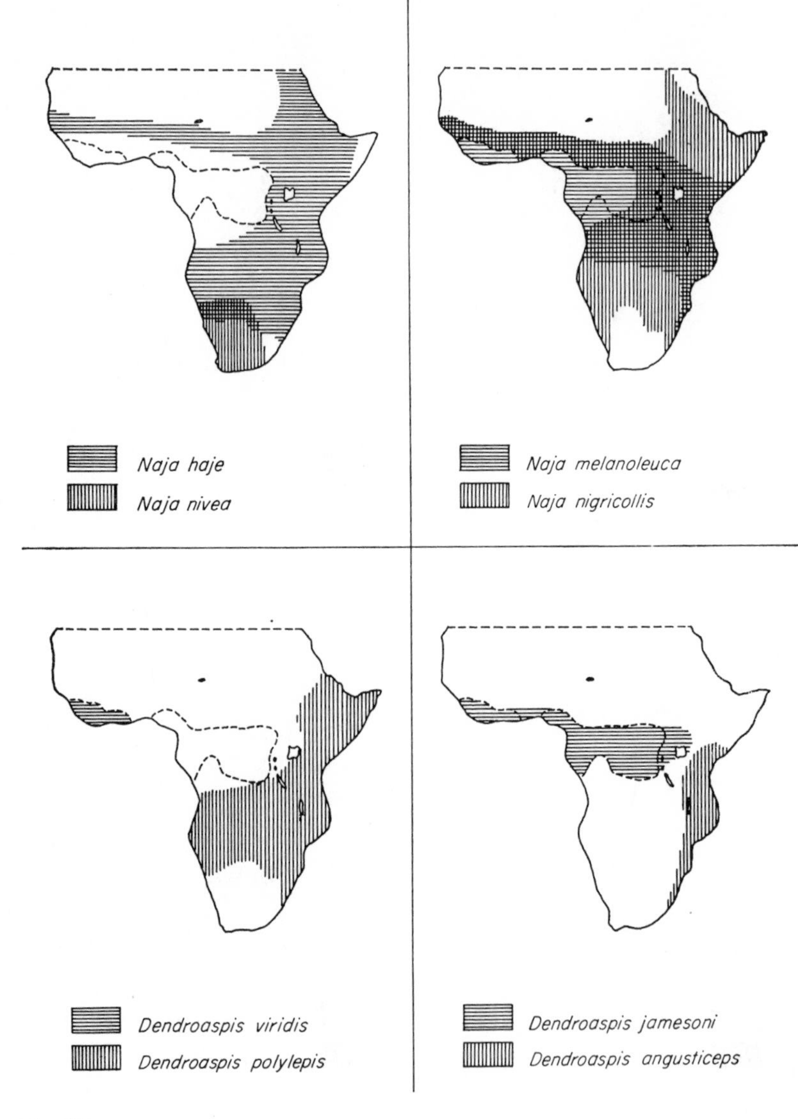
Naja haje
Naja nivea
Naja melanoleuca
Naja nigricollis
Dendroaspis viridis
Dendroaspis polylepis
Dendroaspis jamesoni
Dendroaspis angusticeps

Fig. 37

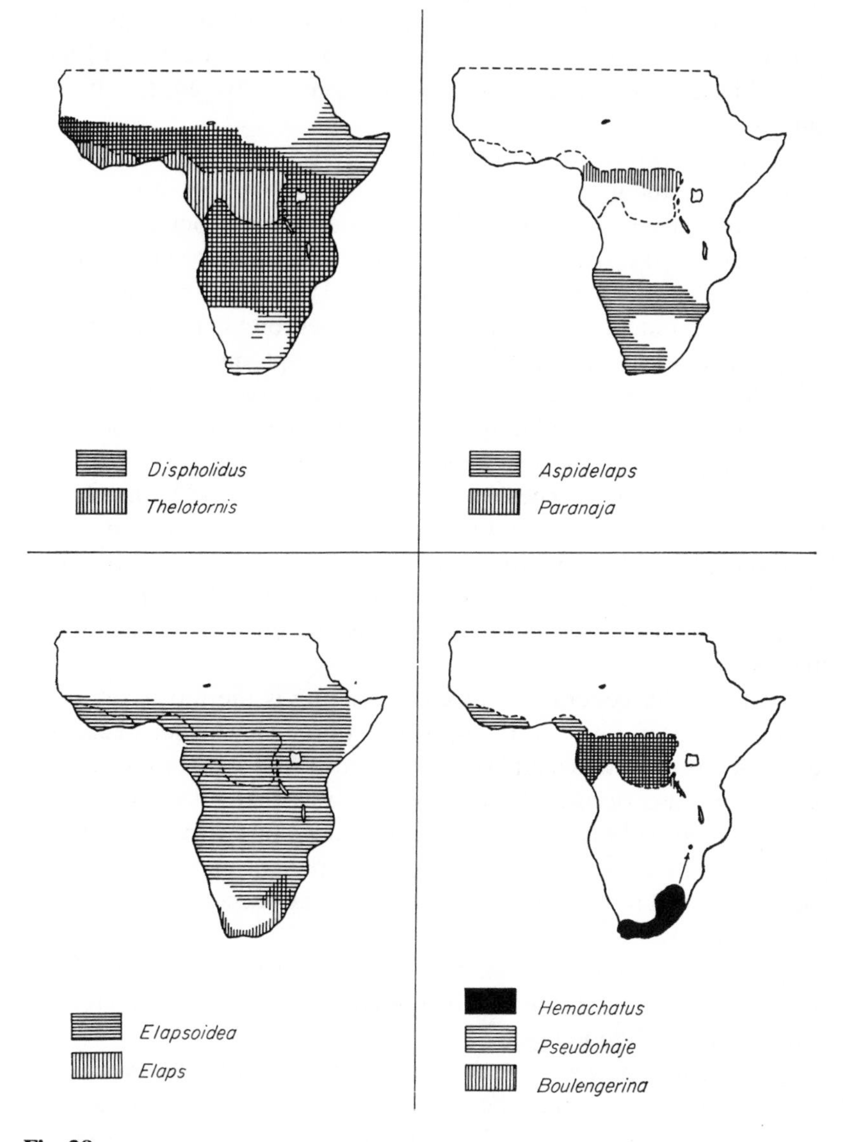
Dispholidus
Thelotornis
Aspidelaps
Paranaja
Elapsoidea
Elaps
Hemachatus
Pseudohaje
Boulengerina

Fig. 38

worst, a slight swelling and anesthesia around the site of the bite. The general symptoms are characteristic for a neurotoxin: vomiting, visual disturbances, abnormal reflexes, convulsions and finally paralysis.

Death occurs from respiratory paralysis. Antivenins are available for treatment.

Naja species. The clinical manifestations of poisoning are very similar for all the African cobras, although a difference may be seen in severity and extent. The dominant factor in severe cases is the neurotoxin, but in many other cases no neurotoxic manifestations have been observed. There are local symptoms that regularly include severe pain, and sometimes a burning pain that frequently radiates up the limb and may last for more than 10 days depending on the size of the local necrosis. Swelling commences 2–3 hours after the bite and reaches its maximum after 24–48 hours. Limb swelling occasionally spreads to the trunk; it may persist for up to 18 days. Bloody blisters frequently develop, and necrosis of the skin and subcutaneous tissue usually appears by the fifth day. These wounds may take several months to heal and in any case have to be treated surgically.

Among the general symptoms, vomiting is the most common, especially after bites of the spitting cobra. Headache, dizziness, unconsciousness, visual and speech disorder, and sometimes convulsions may also occur. The muscle paralysis that seizes the whole body is a particularly severe symptom. It develops within the first 10 hours and may last up to four days. The patient may be unable to open eyes and mouth, or even swallow. Death occurs from respiratory paralysis.

The Asian cobra bite symptoms are very similar to those mentioned above. In addition the cardiotoxin causes sweating, increased heart rate and hypotension.

Bungarus species: Local symptoms are not observed; the neurotoxic symptoms are the same as with *Naja species.* Abdominal pain is found in many cases. Recovery from severe paralysis is rare.

Haemachatus species: Swelling and local pain are slight. Mamba bites are rapidly followed by paralysis and death.

A survey of the symptoms following envenomations by Australian snakes is given in Table 39.

Table 39. Envenomation symptoms of Australian venomous snakes

Species	Local pains	Vomiting	Neurotoxic symptoms	Cardiotoxic symptoms	Peripheric circulation
Notechis scutatus	+	+	Apathy speech disorder swallowing difficulties respiratory paralysis	Myocarditis	Circulatory collapse sweating
Pseudechis porphyriacus	–	+	Minimal	Minimal	Exhaustion
Acantophis antarcticus	–	+	Numbness apathy swallowing difficulties respiratory paralysis	Weakness	Peripheric circulatory failure sweating
Denisonia superba	–	+	Coma	Existing	Peripheric circulatory failure
Denisonia textilis	+	+	Headache numbness dyspnea	Heart failure	Peripheric circulatory failure
Oxyuranus scutellatus	–	+	Paralysis respiratory paralysis	Heart failure	Peripheric circulatory failure

Treatment

Traditional first-aid measures such as suction, incision, excision and ligature are impractical and of no value after elapid bites. The best measures are rest, reassurance and transfer to a hospital. Admission to hospital should be made even in the absence of a wound or of symptoms, especially since in some cases the symptoms are delayed. In any case the patient should be given antivenin as soon as possible. The risk of dying from the bite is greater than the risk of anaphylaxis, and delayed administration may result in reduced efficacy of the antivenin. After antivenin is given, the patient must be observed continuously for at least 48 hours. The route of administration is invariably intravenous. Since the amount of venom deposited is always unknown, the dosage is more or less empirical. The dose of cobra antivenin required to neutralize the average amount of venom is about 300–350 ml. Since complications have to be taken into consideration, facilities must be available to treat them. The best place is the intensive care unit, where the antivenin can be given by an intravenous infusion instead of direct injection.

The antivenin is a foreign protein and may itself produce complications. Indeed they occur in about 50% of all cases. Most of these reactions are mild: itching, urticaria, edema, fever and shivers. If the skin reactions occur, adrenaline or an antihistamine should be injected. About 10% of the cases are serious, and it is essential to monitor blood pressure. Once blood pressure returns to normal, further complications are rare.

Anaphylaxis is observed in about 3% of the cases. The use of drugs such as neostigmine or corticosteroids is controversial and has been of no value in most cases reported so far. The condition of the patient may not be stable during the first 48 hours after the bite, and they are best treated under intensive care. Local symptoms need not be treated initially, but skin and tissue necrosis are treated by excision; skin grafting may be necessary. Antibiotics can be given against secondary infections; Tetanus antitoxin usually is not necessary.

7.2 Hydrophiidae (See Snakes)

The *Hydrophiidae* comprise 13 species of the subfamily *Laticaudinae* and 39 species of the subfamily *Hydrophiinae*. All of them are completely adapted to aquatic life, and they are venomous without exception. Their distribution is limited to the tropical coastal regions of Northeast Africa, Asia (including the islands) and the Pacific coast of Central America (Fig. 39). Although sea snakes are not considered aggressive, many accidents occur in which usually fishermen, rarely bathers, are the victims.

The risk of being bitten by a sea snake is relatively low. According to H. A. Reid, 13 bathers were bitten by sea snakes while bathing in Penang (Malaysia) from 1957–1959; during the same time 140 people were stung by catfish, stingrays or jelly fish while bathing on the same beaches.

The course of an envenomation is mainly dependent on the amount of venom injected. All sea snakes possess in their venom glands many times the amount lethal for humans (usually 10–50 mg of dry, crude toxin can be obtained per animal; the lethal dose for a human is 3–10 mg, depending on the species.) During a bite the sea snake ejects most of its venom in order to kill its prey as fast as possible. It takes a number of days for the venom to be regenerated. Thus whether or not intoxication occurs in a subsequent bite depends on the amount of venom available in the glands, and the symptoms may vary between extremes from "no envenomation" to "fatal". The mortality rate is estimated to about 17% according to a cautious estimate; it may well be higher since the natives of these areas are reluctant to talk about snake bites in general and fatalities in particular, possibly because of fear, suspicion or superstition. The medically important sea snakes are listed in Table 40.

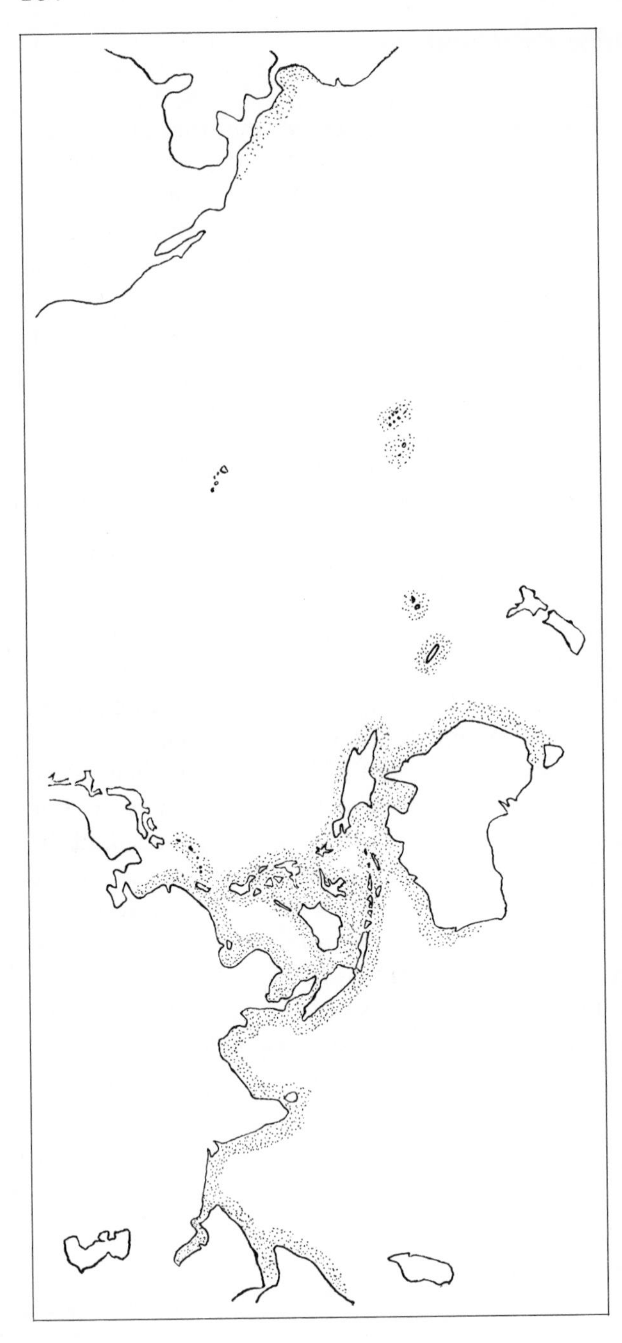

Fig. 39. Distribution of sea snakes *(Hydrophiidae)*

Table 40. Sea snake species and venom yields; Snake and Venom Research Institute, Malaysia, 1960–1963

Species	Number of specimens	*Venom yield (mg dry weight)* Average	Maximum
Enhydrina schistosa	2403	8.5	79
Hydrophis brookei	262	1.1	2
Hydrophis caerulescens	4	–	–
Hydrophis cyanocinctus	1074	8.2	80
Hydrophis klossi	312	1.0	2
Hydrophis melanosoma	4	–	–
Hydrophis spiralis	349	2.1	8
Kerilia jerdoni	188	2.8	5
Lapemis hardwickii	115	1.9	15
Microcephalophis gracilis	19	–	–
Pelamis platurus	3	–	–
Praescutata viperina	2	–	–
Total	4735		

As may be seen from this table *Enhydrina schistosa* is the most common one, and it abounds in most localities on Asian coasts. It is much more aggressive than any other species. A survey of bites from different species may be seen in Table 41. It reflects the cases in one hospital.

Table 41. Fatalities after sea snake bites (H. A. Reid, Epidemiology of Sea-Snake Bites J. Trop. Medicine Hyg. 78, 106–113 (1975). Sea Snake Bites, Br. Med. J. 2, 73 (1956).

Species	No. of bites	No. of fatalities	
E. schistosa	79	14	17.7%
H. cyanocinctus	24	4	16.0
H. spiralis	17	2	11.8
H. klossi	16	1	6.3
K. jerdoni	3	0	–
L. hardwickii	4	1	25.0
M. gracilis	3	1	33.3
total	146	23	15.6

Poisoning

The bite of a sea snake usually produces no pain initially, and other local symptoms such as bleeding, swelling or ecchymosis are not observed. In trivial cases myalgic symptoms may start from $^1/_2$–$3^1/_2$ hours after the bite. Muscle aches, pains and stiffness in the neck, throat, tongue, shoulders, trunk and limbs occur. Myalgia in trivial cases is never severe; and moderate pain is observed 1–4 hours after the bite. These reactions disappear without antivenin in 1–3 days.

Serious cases produce a "thick" feeling of the tongue, thirst, sweating, vomiting and headache. Passive movement of the limb is painful. Blood pressure is normal. Myoglobinuria is evident 3–8 hours after the bite, first as a dusky yellow color, then as a red, brown or black discoloration resulting from the lesions of the muscles. Ptosis is another symptom as well as inability to swallow. The patient is mentally alert, however. Temperature may be slightly elevated.

Fatalities are a consequence of respiratory failure, cardiac arrest or renal failure. Death may occur as soon as three hours after the bite or after 24 hours. 25% of the fatalities occur within 8 hours after the bite, 50% within 8–24 hours and the rest after one day. After 48 hours the chances of survival are good.

Treatment

The victim should be taken immediately to a hospital, and the most important therapy is the injection of antivenin. Recovery is then very rapid. Administration of antivenin can be effective even days after the bite. Other methods of treatment are not indicated; they may temporarily improve the condition of the patient, but there is no real recovery. A suitable dosage is 2–4 ampoules for trivial cases, and 6–20 ampoules in severe cases. Complete recovery may take several months.

Chemistry

The crude venom is a colorless or light yellow viscous liquid from which the mixture of toxins crystallizes in colorless platelets. By means of electrophoresis the crude venom can be separated into five main substances and a number of minor components.

Four compounds are mainly responsible for the symptoms:

1. A neurotoxin or myotoxin, causing the muscle paralysis, is responsible for death.
2. Lecithinase, which causes a lysis of the erythrocytes.
3. Anticoagulase, which prevents the coagulation of the blood plasma.
4. Hyaluronidase, which as diffusion factor is responsible for the rapid spread of toxin in the tissue.

The amino acid sequences of the neurotoxins of *Laticauda semifasciata,* erabutoxins a, b and c, are found in Table 35 as well as the primary structures of the toxins 4 and 5 of *Enhydrina schistosa.*

7.3 Viperidae (Vipers)

Vipers are distributed throughout the Old World; they abound from Europe and the Mediterranean area to Southeast Asia, and they are also native to Africa. The genus *Vipera* is found in Europe (*V. berus, V. aspis, V. ammodytes, V. ursinii, V. kaznakovi, V. latasti, V. palaestinae* in the Eastern Mediterranean area) and Asia (*V. berus, V. ursinii, V. russelli* in India, South China, Burma, Indonesia, Thailand, Sri Lanka and Taiwan); *Echis* and *Bitis* are found in Africa and Arabia.

The mortality rates have been investigated by the WHO. In Sweden about 1,300 bites from *V. berus* are recorded annually;

Table 42. Frequency and mortality of adder bites in Germany

Country	Period	No. of accidents	Lethal bites
German Reich	1883–1892	216	14 (6.4%)
German Reich	1907–1912	256	6 (2.3%)
Prussia	1920–1925	150	1 (0.7%)
Federal Republic	1952	20	2 (10%)
Federal Republic	1964–1969	211	0
Total		853	23 (2.7%)

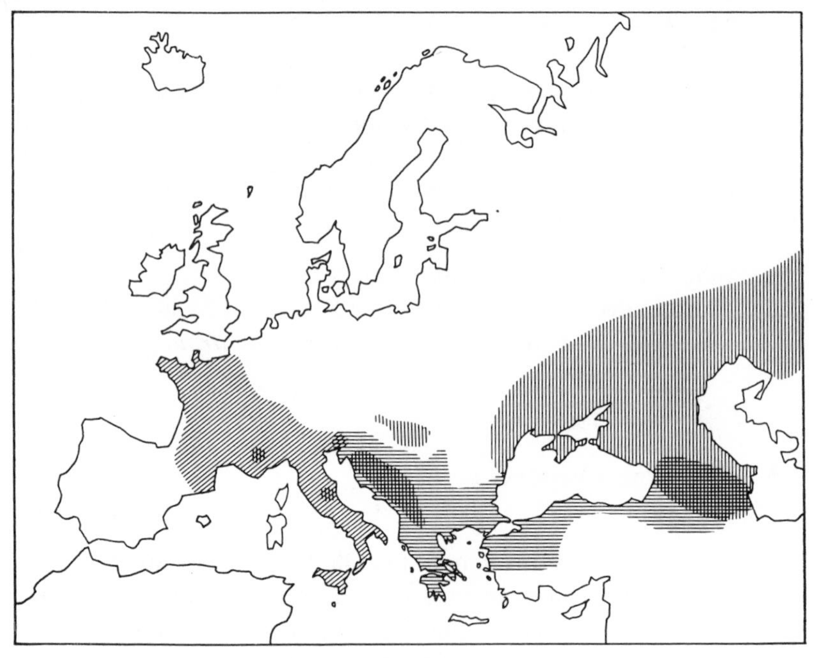

Fig. 40 . Distribution of *Vipera ammodytes*
(Symbols: ≡ *V. aspis* /// *V. ursinii* ||| (according to Klemmer)

12% of them lead to hospitalization. The fatality rate is 0.3%. In the United Kingdom 95 victims have been hospitalized in 100 years, 14 of whom died. (One death was from anaphylactic shock due to subcutaneously injected venom.) A survey of *V. berus* bites in Germany is given in Table 42.

In Israel the number of bites from *V. palaestinae* is between 130 and 260 per year; the fatality rate is 6%.

In France the mortality has been 21.5% in 1944–1947, in Italy 17.6% (1944–1948), in Spain 5.3% (1946–1948) and in Norway 4.4% (1946–1950).

Bites from *Echis carinatus* can be a public health problem since they are rather aggressive. The same is true for the puff adder, *Bitis lachesis*.

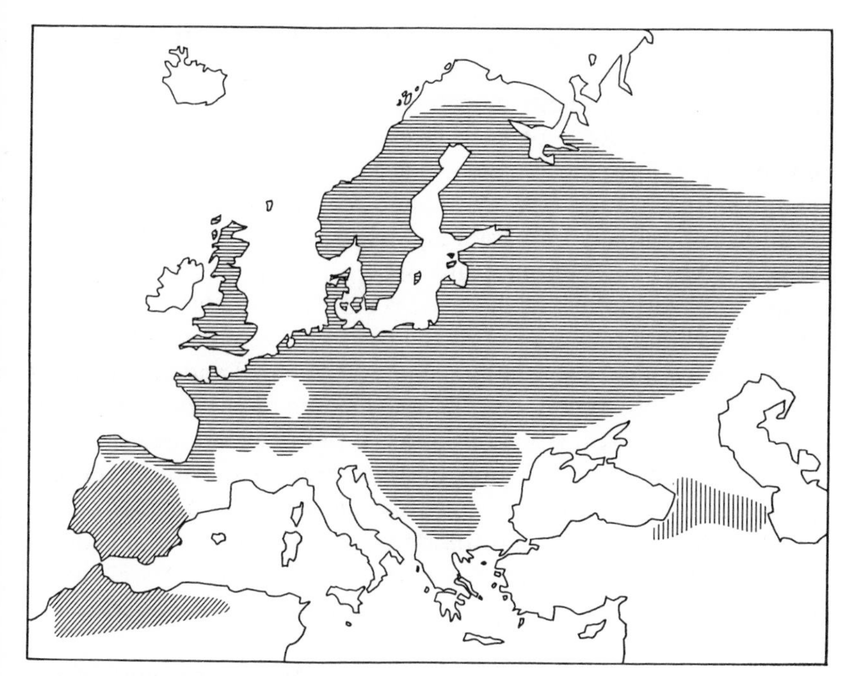

Fig. 41. Distribution of *Vipera berus*
(Symbols: ≡ *V. kaznakovi* /// *V. latasti* ||| (according to Klemmer)

Poisoning

Vipera berus: The bite produces extreme pain and stinging sensations. The site is swollen within a few hours, and an ecchymosis of several centimeters in diameter is observed. Pains finally yield to itching and disappear within 8–10 days. In severe cases vomiting and cardiovascular collapse has been observed. Fatalities only occur in those cases in which blood vessels have been hit.

Vipera aspis: Symptoms closely resemble those of *V. berus,* but are generally more severe. The ecchymosis is larger and may lead to necrosis within a few hours. Generalized symptoms are vomiting, elevated temperature, weak pulse and dizziness. In severe cases renal disorders and damage to the liver are observed. Death occurs very rapidly if a blood vessel was struck.

Vipera ammodytes: This snake is native to Southeast Europe, but is also found in Italy and Austria. These areas are very popular among tourists, and the number of snakebites is increasing in the countryside, coastal areas and offshore islands. Since this snake is a good swimmer, incidents involving bathers occur every year. It is important for victims of such bites to keep a clear head, return immediately to the beach and see a doctor. Most of these envenomations are harmless and there is no reason for panic. The bite is painless and initially only the fang marks, which usually bleed, are noticed. After 1/2–1 hour pains arise, the limb becomes red and swollen, while the rest of the body is pallid. Frequently ecchymosis develops within one hour after the bite. In severe cases the lymph nodes are swollen, the victim becomes frightened and an increased pulse rate is observed as well as short, labored breathing. Collapse is not uncommon. After several hours, a red or violet area of tissue hemorrhage develops around the bite area, and without proper treatment necrosis occurs, which usually leads to the loss of the afflicted limb. Death from respiratory paralysis or from brain edema occurs if the venom is injected directly into a blood vessel. In all cases the victim should be treated with antivenin, if possible, in a hospital.

Vipera palaestinae: Immediately after the bite, swelling of the limb arises, which can become widespread in hours and days. The swelling is accompanied by hemorrhage, hematomas and the formation of blisters at the bite site. General symptoms are nausea, vomiting, abdominal pain, diarrhea and peripheral shock. Occassionally fever and anemia have been observed. The victim will usually recover within one week, but even after this time fatal shock may occur. In such cases the patient becomes uneasy, and sweats; breathing is rapid and body temperature rises. Tachycardia is associated with a fall of blood pressure down to shock values. If treated properly the patient can be saved, but if a blood vessel was struck by the bite death occurs within a short time.

Vipera russelli: Symptoms closely resemble those of V. palaestinae. Cause of death is a hemorrhagic diathesis.

Bitis spp.: Local pain and swelling are noticed within 20 minutes after the bite. The swelling reaches a maximum one-two days after the bite, and it takes five days to three weeks to resolve. Local blistering and necrosis are found in 50% of the cases. Systemic envenomation includes spontaneous bleeding with thrombocytopenia, hypotension and bradycardia. Vomiting, abdominal colic, dizziness and drowsiness are common as well as hyperpyrexia, sweating and urticaria. Death may be due to renal failure or circulatory collapse.

Echis carinatus: The symptoms are similar to those of *Bitis spp.* The mortality rate is rather high; even in treated cases it may reach 6–16% depending on the season. It has been observed in Nigeria that the rate in December and January is three times as high as in the other months. Swelling starts within a few minutes after the bite and may continue to increase for 72 hours. Discoloration and blisters around the bite site are common, and there is continued oozing from the fang marks. Hemorrhage, usually accompanied by abnormal clotting is the outstanding symptom of *Echis* poisoning. Bleeding into the brain or other vital organs usually is the cause of death.

Treatment

Although already mentioned above, treatment by cauterization, incision or suction is completely useless; on the contrary they are dangerous for the patient. Victims with local symptoms should only be treated symptomatically, if at all. The use of serum should be limited to those cases in which symptoms of systemic envenomation are observed, such as vomiting, diarrhea and hypotension soon after the bite. Patients treated by serum usually improve rapidly. The antivenin must be specific and highly purified, and given in slow intravenous injections; usually 50–80 ml is an adequate quantity. It is important to consider that *Vipera russelli* and *Echis spp.* have developed "geographic" races or subspecies that differ considerably in their toxins, so that antivenin derived from another subspecies may be ineffective.

7.4 Crotalidae (Pit Vipers, Rattlesnakes)

Pit vipers and rattlesnakes abound in the New and Old World. In both North and South America they are the dominant element among venomous snakes. According to Russell, 45,000 people in the United States are bitten by snakes each year, 8000 of them by venomous snakes, and 7000 develop signs of envenomation. Between 12–15 victims die. Almost all of these deaths are attributed to rattlesnakes, which account for about 60% of all bites from venomous snakes in the US. Most remaining accidents are attributed to the other crotalids, copperhead and cottonmouth; coral snakes inflict about 1% of the bites.

Rattlesnakes derive their name from the last joint of their tail. If they feel threatened, they use it to generate a rattling noise. They attack only if this behavior does not chase away the enemy. The name pit viper comes from the two small pits on the head between the eyes and nostrils. These are temperature sensitive organs (a sort of infrared eye) by means of which the snake may differentiate between warm-blooded animals and other moving things.

North America: Most venomous bites are inflicted by *Crotalus adamanteus, C. atrox, C. horridus, C. durissus, C. viridis, Agkistrodon contortrix* and *Sistrurus catenatus.* In colloquial language, "rattlesnakes" are differentiated from "mocassins". This classification corresponds to the genera *Crotalus* and *Agkistrodon.* A summary of the most important North American species is given in Table 43. Table 44 reflects the symptoms of envenomation from both of these genera.

South America: 11 of the 13 South American *Crotalidae* belong to the genus *Bothrops;* the two others are *Crotalus durissus terrificus* and *Lachesis muta muta* (bush master). *C. durissus terrificus* accounts for 10% of the bites, the figure for *L. muta muta* is minimal, 0.2%. All of the remaining envenomations are caused by *Bothrops spp. B. jararaca* (52%), *B. jararacussu* (10%), *B. alternatus* (6%) and *B. neuwiedii* (4%) are the most important ones. The high percentage of bites by *B. jararacussu* is striking if one considers its rather limited geographic distribution. This reflects the aggressiveness of the species; the natives in the interior fear this snake more than anything else.

Asia: Two genera account for most of the snakebites, *Agkistrodon* and *Trimeresurus. Agkistrodon halys* is distributed from the European part of the Soviet Union to Siberia, Tibet, China, Korea, Taiwan and the Riu-Kiu Islands; in Japan *Agkistrodon halys blomhoffii,* the mamushi, is well known. *Agkistrodon rhodostoma,* the Malayan pit viper, abounds in South East Asia. *Trimeresurus spp.* are also responsible for many bites in Afghanistan, Pakistan, India, Sri Lanka, Kashmir, Nepal, Tibet, China, Korea, Japan,

Fig. 42. Distribution of *Crotalus terrificus* (according to Amaral, Machado, Marcelo Silva Jr. from W. Bücherl in Behringwerke Mitt., Die Giftschlangen der Erde, Marburg, 1963)

Fig. 43. Distribution of *Lachesis muta, Bothrops alternata, Bothrops ammodytoides* (partially from Machado and Marcelo Silva Jr. from W. Bücherl in Behringwerke Mitt., Die Giftschlangen der Erde, Marburg 1963)

Fig. 44. Distribution of *Bothrops atrox, B. cotiara, B. jararaca* (from W. Bücherl in Behringwerke Mitt., Die Giftschlangen der Erde, Marburg, 1963)

Taiwan, Burma, Thailand, Vietnam, Laos, Cambodia, Malaysia and the Indo-Australian Archipelago including the Philippines. On the island of Okinawa alone 300 bites from *T. flavoviridis,* the Habu, are recorded each year.

Poisoning

Crotalus spp.: In *Crotalus* envenomation the first four hours are the most critical ones. Delayed or inadequate treatment may lead to tragic consequences. Envenomations by rattlesnakes will usually show diagnostic symptoms within the first ten minutes. In most patients significant swelling will be seen in the first moments. The swelling may be severe eight hours after the bite. Immediate pain is common. Ecchymosis and discoloration at the bite site appear within several hours in untreated cases. Hemorrhagic vesiculations and petechiae are common if insufficient antivenin is given, and in severe cases serious tissue necrosis develops.

The general symptoms are paresthesia around the mouth and muscle fasciculations, both of which are early findings after the bite. Weakness, sweating, faintness and nausea are commonly reported after rattlesnake bites. The skin temperature over the affected part is elevated; regional lymph nodes may be enlarged, painful and tender.

In severe cases bleeding phenomena may be observed such as hemmorrhage from the gums, hematemesis, melena and hematuria. Respiratory distress may develop and respiratory failure ensue. In fatal situations, shock leading to complete cardiovascular failure usually precedes death. In South American *Crotalidae,* e. g. *Crotalus durissus terrificus,* the local symptoms mentioned above are much less severe; usually only a local paresthesia is observed.

Bothrops spp.: After the painful bite an ecchymosis develops around the bite site. The venom of *B. jararacussu* also causes hemorrhage and bleeding, not only in the limb involved but also in other parts of the body. Local swelling, blisters and necrosis may lead to severe tissue damage if not properly treated. The systemic symptoms are: dizziness, nausea and vomiting with bloody expectoration. Pulse and temperature are usually normal. A lowering of the body temperature and decreased blood pressure indicates the onset of shock; increased temperature indicates the starting of necrosis or secondary infections.

Agkistrodon rhodostoma, A. halys: The bite is usually but not always painful. The first symptom is swelling of the site concerned, which may last for three days. Blisters around the bite site are common, and in about 10% of all cases a local necrosis develops. A typical symptom is hemorrhage with total, 2–3 weeks lasting, afibrinogenemia. 20 minutes after the bite blood may already occur in the sputum, as the lungs are affected first. Hemorrhage into the brain or peritoneum may lead to death. Apathy, strong and lasting thirst, increased pulse frequency and lowered blood pressure are characteristics. Death usually occurs relatively late, usually not before the third day, but sometimes after one week. There is time enough for proper treatment.

Trimeresurus spp.: Trimeresurus bites progress similarly to those of *Agkistrodon spp.,* but usually they are less severe. In most cases only the local symptoms are observed. Fatalities are rare.

Treatment

The amount of polyvalent (Wyeth) antivenin *(crotalidae)* to be given (excluding copperhead for which antivenin is rarely needed) depends on the severity of the envenomation. No less than three vials of antivenin should be given intravenously in minimal envenomations with local effects only. In patients with systemic symptoms 5–12 vials, and in severe cases up to 40 vials may be necessary. Measures for combatting shock should be readily available, and in severe poisonings equipment for endotracheal intubation or tracheostomy and a positive pressure breathing apparatus should be in readiness. Broad-spectrum antibiotics may be given if there is severe tissue necrosis, and analgesics may be given for pain.

At the first signs of respiratory distress oxygen must be given. Corticosteroids are not effective in treatment of venom shock states.

Lesions of the skin should be cleansed and sterile covers used. Surgical debridement of blebs, bloody vesicles and superficial necrosis should be performed between the 3rd and the 10th day. Fasciotomy should be discouraged. Although local swelling is a constant feature of most viper bites, it resolves spontaneously, provided there is no local necrosis. Using surgery insures that some necrosis will develop in all victims.

7.5 Colubridae

Among the snakes usually considered "non-venomous" because they do not exhibit the typical venom apparatus with venom fangs) many species have developed a gland homologous to the venom gland of venomous snakes: The Duvernoy gland. Moreover the colubrides possess venom fangs in the back of the throat at the base of which the exits of the Duvernoy's glands end. Because of this anatomic peculiarity colubrides may use a "venom bite" only with small animals. The bites of these animals will endanger humans only if a finger were struck and taken deep into the throat where venom could be injected.

In most species only local reactions are observed, such as swelling, reddishness and paresthesia about the bite site; they disappear in untreated cases within a few days. Among these snakes are *Boiga irregularis, Cerberus rhynchops, Enhydris polylepis, E. punctata, Fordonia leucobalia, Myron richardsoni,* native to Northern Australia, New Guinea and the Solomon Islands. They present no danger to humans.

More serious are the sequela with *Rhabdophis tigrinus* (Japan, Korea, China); after bites of this snake blood clotting disturbances appear, which may persist for a week.

The two only species of venomous colubrides that are able to cause fatalities in humans are *Dispholidus typus* and *Thelotornis kirtlandii.* Both are native to the southern parts of Africa below a line extending from Senegal to Ethiopia (Fig. 38). They do not inhabit the deserts, rather they usually live in trees, which explains their name "boomslang" in Afrikaans. They feed upon smaller reptiles, frogs and the eggs of birds.

They react with fear to humans; they only bite in case of danger or provocation. From 1915 to 1976, 30 serious cases were reported, eight of them fatal (six fatalities from *D. typus* and two fatalities from *Th. kirtlandii*). The real number, however, is probably higher, since in all of these areas no statistics are kept.

Poisoning

The bite is associated with serious, strong subcutaneous bleeding about the bite site, from wounds and from the gums. After a few hours the patient loses consciousness, and renal failure is observed.

Death occurs after several days. Autopsy reveals extensive bleeding in all internal organs as well as fibrin thrombes in the capillary vessels.

Treatment

The only effective therapy is the injection of monovalent antivenin, which until recently had been produced by the South African Institute for Medical Research. The production, however, had to be given up and it is unlikely that ampoules are left over anywhere in the world. Thus there is presently no effective means available for treatment.

References

Bücherl, W., Buckley, E. (Eds.): Venomous Animals and Their Venoms, Vol. 1 (1968) and Vol. 3 (1971). New York, Academic Press

Lee, C. Y., Onyang, C., Chang, C. C.: Snake Venoms, Taipei: National Taiwan University 1974

Kaiser, E.: Tier- und Pflanzengifte. Munich: Goldmann 1973

Vries, A. de, Kochva, E.: Toxins of Animal and Plant Origin. London: Gordon and Breach, 1973

Yang, C. C.: Toxicon *12*, 1 (1974)

Reid, H. A.: Tropical Doctor *2*, 155 (1972)

Russell, F. E.: J. American Medical Association 233, 341 (1975)

Russell, F. E.: Experientia (Basel) *30*, 8 (1974)

Warrell, D. A.: Acta Tropica (Basel) 33, 312 (1976)

Minton, S. A., Minton, M. R.: Venomous Reptiles. New York: Scribner's Sons 1969

U. S. Navy Bureau of Medicine and Surgery: Poisonous snakes of the World. Washington: U. S. Gov't Printing Office 1968

Russell, F. E., Saunders, P. R.: Animal Toxins. Oxford: Pergamon Press 1967

Russell, F. E.: Snake Venom Poisoning, Philadelphia, 1980, J. P. Lippincott

Reid, H. A.: J. Trop. Med. Hyg. *78*, 106 (1975)

Behringwerk-Mitteilungen: Die Giftschlangen der Erde, Marburg 1963.

Klemmer, K.: In Venomous Animals and Their Venoms (W. Bücherl, E. Buckley and V. Deulofeu, Eds.), Vol. I, 1968, Academie Press, New York.

Compilation of amino acid sequences of snake venom toxins

Mebs, D.: Zentrum der Rechtsmedizin, University of Frankfurt, Frankfurt/Main Germany, August 1979

7.6 Heloderma (Gila Monsters)

The only venomous reptiles other than the snakes are the two species of *Heloderma,* that is *H. suspectum* and *H. horridum. H. suspectum* may be up to 50 cm in length. They are native to the Gila River Valley in Arizona, Northwest Arizona, New Mexico and the Sonora desert in Mexico. The bigger species, up to 80 cm in length, inhabits Central Mexico and the Northern parts of Central America.

The animals are fearful and warn by opening their mouth wide and by hissing. They only bite if one attempts to seize them. The bite is very unpleasant as they possess very strong jaws that enable them to clamp down for 10–15 minutes. The muscles are so strong that the mouth can only be opened by application of force, for example with a screwdriver. During this time enough of the venomous saliva may intrude into the wound to produce envenomation.

Poisoning

Bleeding from the usually large wound begins after about 10 seconds and soon becomes strong. The site is swollen; the pains depend on the size of the wound; sometimes discoloration (red or blue) is observed.

The general symptoms are not very distinct. Nausea, slightly elevated pulse (88–92) and low blood pressure are typical, as well as elevation of the temperature up to 39.5 degrees C.

Although the LD_{50} of the crude venom amounts to 1.5 mg/kg (mouse), fatalities in humans have not been observed. The few "fatal cases" that are reported have been the sequel of further effects: Heart failure from fear, abuse of alcohol and secondary infections of the untreated wound.

Treatment

The most important consideration is treatment of the wound, which sould be supported by broad spectrum antibiotics. Beyond this, symptomatic treatment suffices. The patient recovers after a few days up to two weeks.

The occasionally recommended application of polyvalent snake serum is not only useless (*Heloderma* toxin is completely different from all other snake venoms) but dangerous and should not be used.

References

Mebs, D.: Salamandra *6*, 135 (1970)
Mebs, D.: Hoppe Seylers Z. Physiol. Chemistry *350*, 821 (1969)
Mebs, D.: Naturwissenschaften *54*, 494 (1967)
Roller, J. A.: Clinical Toxicology *10*, 423 (1977)

Table 43. Review of the most important North American venomous snakes and their toxicity.

Species	Common English name	LD_{50}mg/kg (i. p.) mouse	LD_{50}mg/kg (i. v.) mouse
A) Crotalidae	Rattlesnakes, pit vipers		
Agkistrodon contortrix	Copperhead	10.50	10.92
A. piscivorus	Cottonmouth	5.11	4.00
Crotalus cerastes	Sidewinder	4.00	
C. willardi	Ridge-nosed rattlesnake		
C. molossus	Norther blacktailed rattlesnake		
C. pricei	Western twinspotted rattlesnake		
C. mitchelli	Southwestern speckled rattlesnake		
C. tigris	Tiger rattlesnake		
C. lepidus lepidus	Mottled rock rattlesnake	L	
C. ruber ruber	Red diamondback	6.69	3.70
C. atrox	Western diamondback	3.71	4.20
C. adamanteus	Eastern diamondback	1.89	1.68
C. scutulatus	Mojave rattlesnake	0.23	0.21
C. enyo enyo	Lower California rattlesnake		
C. viridis viridis	Prairie rattlesnake	2.25	1.61
C. v. oreganus	Northern pacific rattlesnake		
C. v. helleri	Southern pacific rattlesnake	1.60	1.29
C. v. lutosus	Great basin rattlesnake	2.20	
C. v. cerberus	Arizona black rattlesnake		
C. horridus horridus	Timber rattlesnake	2.91	2.63
C. h. artricaudatus	Canebrake rattlesnake		
Sistrurus catenatus catenatus	Eastern massasauga	4.00	3.70
S. c. tergeminus	Western massasauga		

Table 43 (continued)

Species	Common English name	LD_{50}mg/kg (i. p.) mouse	LD_{50}mg/kg (i. v.) mouse
S. c. edwardsi	Desert massasauga		
S. miliarius miliarius	Pigmy rattlesnake	6.70	4.01
S. m. barbouri	Pigmy rattlesnake		
S. m. streckeri	Pigmy rattlesnake		
B) Elapidae			
Micruroides euryxanthus	Arizona coral snake		
Micrurus fulvius	Eastern coral snake		

Table 44. Comparative review of the symptoms of envenomation by North American venomous snakes

Symptoms	Rattle-snakes *Crotalus spp.*	Mocassins *Agkistrodon spp.*	Coral snakes *Micrurus spp.*
Local swelling, edema	+++	++	+
Local pains	++	+	++
Discoloration of skin	+++	+	+
Blisters	+++	+	−
Ekchymosis	+++	++	−
Superficial thrombosis	++	−	−
Necrosis	++	+	−
Scabling of tissue	++	−	−
Burning, itching of skin and numbness[a]	++	+	+++
Fasciculation, muscle convulsion	+	−	+
Muscle weakness, paralysis	+	−	+
Cramps	−	−	−
General weakness	++	+	+++
Thirst	++	+	−
Nausea, vomiting	++	+	+
Diarrhea	+	−	−
Headache	−	−	−
Abdominal pain	−	−	+
Excessive salivation	−	−	+++
Glycosuria	++	+	
Proteinuria	++	+	
Weak or irregular pulse	+++	+	++
Hypotension, shock	+++	+	+

Symptoms	Rattle-snakes *Crotalus spp.*	Mocassins *Agkistrodon spp.*	Coral snakes *Micrurus spp.*
Destruction of erythrocytes	+++	−	
Prolonged bleeding time	++	−	
Elongated coagulation time	+++	−	
Hemorrhage[b]	+++	+	
Anemia	++	−	
Changing of blood laminas	++	−	
Ptosis	+	−	++
Visual disturbances	+	−	++
Respiratory difficulties	++	+	++
Speech disorder, difficulty of swallowing	−	−	++
Swelling of regional lymph nodes	++	+	+
Anomalous ECK	+	−	−
Coma	+	−	−

+, ++, +++: degree of severity of the symptoms; in severe cases the individual symptoms may be stronger than normally observed.

[a] Often limited to tongue and mouth, but may also involve the scalp, toes and fingertips.

[b] The bleedings may occur in the gastrointestinal tract, in the urinary passages, in the lungs or at the gingiva: hemorrhage from gums is generally observed with *Bothrops* bites.

8 Therapeutic Use of Animal Venoms

Throughout history there were many attempts to use animal venoms in medicine. Mythology abounds with references to venomous animals, which were frequently attribute of gods or medical tools. For example, rattlesnakes were the attributed of the Aztec goddess Coatlicue (Mother of Gods). The staffs of Indian medicine men are frequently decorated with snakes, as was the serpent-entwined staff of Asklepios, God of Medicine in Ancient Greece. The Greeks and Romans had houses in which Esculap adders, *Elaphe longissima*, were held in honor of Asklepios. This non-venomous snake was brought by the Romans from the Mediterranean Sea to other parts of their empire for this purpose, and even today populations of *Elaphe longissima* can be found in some places in Austria as well as near the town of Schlangenbad (close to Wiesbaden) in Germany: They date from these times.

8.1 Snake Venoms

Since the beginning of the twentieth century there have been investigations directed toward the utilization of snake venoms in therapy. Unfortunately, some unreliable findings made headlines and raised hopes that could not be fulfilled. Ambiguous and contradictory results and certainly many failures and disappointments originate from experiments with inexactly classified snakes or from the use of crude venoms very complex and varied in composition. Composition may vary seasonally and locally, and thus experiments are useful only if performed with purified single components or with preparations meeting certain standards. Those positive results obtained so far are limited to analgesic action and influence on blood clotting.

Analgesic Activity

The analgesic effect of snake venoms is of some importance in cases of terminal cancer because of the extreme pains usually requiring the use of morphine. Studies have been made on the toxins of *Vipera berus, Crotalus adamanteus, C. durissus terrificus, Bothrops alternatus, B. cotiara, B. jararaca, B. jararacussu* and *Naja naja*. The findings were controversial, but best results were obtained with cobrotoxin *(Naja naja)*, which has turned out to be an important and useful alternative to morphine since its analgesic activity is higher.

In cases of arthritis and rheumatism, toxins from *V. aspis* and *V. ammodytes* produced satisfactory results. Cobrotoxin, too, has been applied in polyarthritides and *Cortalus* toxin against migraine. Positive results were obtained in 50–70% of the cases. Obviously these divergent results have prevented further application.

Effects on Blood Clotting

Two different effects have to be considered: inhibition of blood clotting and enhancement.

The use of snake venoms as anticoagulases, ie, for preventing blood clotting, goes back to an observation by A. H. Reid. The venom of *A. rhodostoma* contains a thrombolytic factor that can be isolated in pure form and has been named Ancrod (WHO); registered trade marks are *Arwin* and *Venacil*. Blood clots can be dissolved with this preparation, and thrombophlebitis as well as vascular clogging can be treated successfully. Arwin has been used with good success in cases of thrombosis of veins and arteries, chronic arterial disturbances of blood supply and some cases of Angina pectoris.

D. von Klobusitzky isolated clot-promoting fractions from *Bothrops artrox* and *B. jararaca* venoms, and already in 1936 he suggested that this substance could possibly be used for therapeutic purpose in the treatment of haemophilic patients. Since then intensive studies have been performed; they resulted in a preparation that has been known for a few years under the trade mark *Reptilase* (from *B. atrox*). It has proved to be a very effective styptic without side-effects. Application may be intravenous, subcutaneous or intramuscular. It is mainly used prophylactically as a hemostatic agent in connection with surgical interventions such as tonsillectomy, prostatectomy and plastic surgery. It has also been used, however, in the treatment of bleeding (nose, lungs, carcinoma, gynaecologic) or of hemorrhagic diathesis.

Other Applications

Up to 1960 preparations from *Crotalus* venom *(C. durissus terrificus)* were used in the treatment of epilepsy, with good success in 75% of the patients. In recent years these preparations have been replaced by other psychopharmaca.

8.2 Bee Venom

Bee venom preparations (e. g. Forapin "Mack") have been successfully used in treatment of rheumatic diseases, pains of muscles and joints, arthritis, neuralgia, frostbite and muscle strains. Good results have also been obtained in cases of chronic polyarthritis.

The activity of bee venom has been well known since ancient times. The route of application at that time, however, – the bee sting – was neither pleasant nor harmless.

It was K. Forster who succeeded in 1938 in preparing an ointment in which purified and standardized bee venom is used.

8.3 Toad Venom

Probably the first and oldest therapeutically used animal venom is the skin gland secretion of the toads. Dried and powdered toad skins are used even today in East and Southeast Asia, for example in Taiwan, for treatment of heart diseases and dropsy.

References

De Klobusitzky, D.: In "Venomous Animals and Their Venoms", Vol. III, p. 443 New York: Academic Press 1971
Ehrly, A. M.: Med. Welt *26*, 446 (1975)
Hess, H., Keil-Kuri, E., Marshall, M.: Münch. med. Wschr. *117*, 1317 (1975)
Folia Angiologica, Vol. XXIII, book 10 (1975) (10 essays about Arwin)
Reid, H. A. : Thrombos-Diathes. haemorrh (Stuttg.) (Suppl.) *38*, 75 (1970)
Seegers, W. H. and Ouyany, C.: In Snake Venoms (Chen-Yuan Lee, Ed.), Springer, Berlin, 1979.

Subject Index

List of Institutes Which Produce Antivenins

Against snake bites:

Algeria	Institut Pasteur d'Algeria, Rue Docteur Laveran, Algiers
Australia	Commonwealth Serum Laboratories, Parkville, Melbourne
Brazil	Instituto Butantan, Caixa Postal 65, Sao Paulo
France	Institut Pasteur, Service de Serotherapie, 36 Rue du Docteur Roux, Paris XV
Germany	Behringwerke AG, Postschließfach 167, 355 Marburg
India	a) Central Research Institute, Kausali, R. I., Punjab b) Haffkine Institute, Parel, Bombay 12
Indonesia	Perusahaan Negara Bio Farma, 9 Djalan Pasteur, Bandung
Iran	Institut d'Etat des Serums et Vaccins Razi, Boite Postale 656, Teheran
Israel	Rogoff Wellcome Res. Laboratory, Beilinson Hospital, P. O. Box 85, Petah Tikva
Japan	Institute for Infectious Diseases, University of Tokyo, Shiba Shirokanedaimachi, Minato-Ku, Tokyo
Yugoslavia	Institute of Immunology, Serum Institute, Rockefellerova 2, Zagreb
South Africa	South African Institute for Medical Research, P. O. Box 1038, Johannesburg
Taiwan	Taiwan Serum Vaccine Laboratory, 130 Fuh-lin Road, Shiling, Taipei
Thailand	Queen Saovabha Memorial Institute, Bangkok
United States	Wyeth Inc., Box 8299, Philadelphia 1, Pa.

Against scorpion stings:

Algeria	Institut Pasteur d'Algeria, Rue Docteur Laveran, Algiers
Brazil	Instituto Butantan, Caixa Postal 65, Sao Paulo
South Africa	South African Institute for Medical Research, P. O. Box 1038, Johannesburg

Against spider bites:

Australia	Commonwealth Serum Laboratories, Parkville, Melbourne
Brazil	Instituto Butantan, Caixa Postal 65, Sao Paulo
South Africa	South African Institute for Medical Research, P. O. Box 1038, Johannesburg

Against coelenterate stings:

Australia	Commonwealth Serum Laboratories, Parkville, Melbourne

Against stone fishes:

Australia	Commonwealth Serum Laboratories, Parkville, Melbourne

Against weaver fishes:

Yugoslavia	Medicinski Centar, Pula